COBBETT'S COUNTRY BOOK

COBBETT'S COUNTRY BOOK

An Anthology of William Cobbett's Writings
on Country Matters

Selected and with an Introduction
by
Richard Ingrams

Illustrated by Bert Kitchen

DAVID & CHARLES

NEWTON ABBOT   LONDON

NORTH POMFRET (VT) VANCOUVER

ISBN 0 7153 6800 1
Library of Congress Catalog Card Number

Set in 11 on 13 pt Baskerville by Wordsworth Typesetting Ltd
and printed in Great Britain by Redwood Burn Ltd,
Trowbridge and Esher, for David & Charles (Holdings) Limited,
South Devon House, Newton Abbott, Devon

Published in the United States of America
by David & Charles Inc. North Pomfret
Vermont 05053 USA

Published in Canada by Douglas David & Charles Limited
3645 McKechnie Drive West Vancouver BC

CONTENTS

WILLIAM COBBETT   7

INTRODUCTORY OBSERVATIONS FROM
TREATISE ON COBBETT'S CORN   21

GARDENING   27
The Situation   29
Box Edging   34
The Greenhouse   37
Some Herbs and Vegetables   41
Garden Seeds   64
Trees and Tree Planting   71
Some Trees   74
Pests   88
Gardeners   100
A Kalendar of Work   101

FARMING   119
Cobbett's Farm   121
Cobbett's Sow   133
Ploughing   134
Wheat Growing   143
The Mill   149
Making Bread   150
Keeping Pigs   154
Fowls   166
Pigeons   170
Rabbits   171
Goats   174
Brewing Beer   176

RURAL SPORTS   185
The Game   187
In Defence of Blood Sports   193
Single Stick   198
Hare Coursing   200
Hare Hunting   202

TWO SWEETCORN RECIPES   203

Mush   205

Samp   210

SOURCES   216

WILLIAM COBBETT

Oh, for one man who should write healthy hearty, straightforward English! Oh, for Cobbett!

Hilaire Belloc

WILLIAM COBBETT
1763-1835

William Cobbett is the great forgotten man of English literature. Journalist, agitator, historian and countryman, he wrote prolifically all his life with an absolute mastery of the English language. But today only his best-known book, *Rural Rides*, is still in print.

Cobbett himself is partly to blame for this state of affairs. Although he was a great writer, he was entirely lacking in literary interests. His approach to everything was practical, and he used the English language to instruct, exhort and abuse, but never to please. He had no concern with form. He wrote whatever came into his head and his writing is an apparently endless 'stream of consciousness' which does not fit naturally into finite books. Even when Cobbett sat down to write a book on a specific topic he could not stop himself digressing. 'Thoughts come,' he said, 'much faster than we can put them upon paper.' But Cobbett did his best to catch up with them. Writing became almost another instinctive human activity to him, like walking or sleeping. The quantity of his output has been exceeded only by Defoe.

The result is that unlike most other writers Cobbett is an ideal subject for an anthologist. The only difficulty is in knowing where to begin. It would be possible to compile a handful of collections all of equal merit. The present volume confines itself to Cobbett's writings on Country matters.

No one deserves the description 'radical' more than Cobbett. He went to the roots of any matter, in the case of agriculture and farming, literally. These activities were not a sideline to him, a relaxation from journalism and politics. He expressed his personality as vividly in his writing on country matters as he did in his *Register*. In both his attitude is equally intense. He approaches, for example, with the same relished hostility the pests and vermin which threaten the farmer as he does his political and historical opponents. The garden or the farm is a microcosm in which Cobbett's role is, as always, to promote order and good sense in defiance of a host of enemies and false theorists. It would be wrong therefore to regard these passages as a charming and overlooked backwater of Cobbett's work. They are as good and as typical of him as anything else he wrote.

Cobbett's life reads like an old-fashioned romance. He was born just over 200 years ago in 1763, the third of four sons of George Cobbett, a Hampshire farmer, and his wife Ann. From his earliest years he worked on the farm, as a human scarecrow in a little blue smock, barely big enough to climb the gates and stiles. Later, when he was older, he weeded the wheat and led the horse for harrowing barley, and eventually was permitted to reap and plough. His father taught him his three Rs, but apart from this he had no formal education. At the age of eleven he got a job working in the Bishop of Winchester's garden at Farnham.

Three times he left home, apparently from no other motive but the love of adventure. The first time, inspired by what a fellow gardener had told him of their splendour, he set off for Kew Gardens, with 'thirteen half-pence' in his pocket. He was given a job at Kew, for how long is not certain. At the age of twenty-one he left home again to stay with his uncle in Portsmouth. Here the sight of the fleet anchored off Spithead was too much for him and he decided to run away to sea. The captain of the man-of-war where he applied, however, assumed quite wrongly that Cobbett was escaping from a shotgun wedding and refused to take him on board.

In 1783 Cobbett left home for good. He set out on a May morning, dressed in his holiday clothes, to escort three girls to Guildford fair but when he came to cross the main road he saw the London stage coach rattling down the hill towards him. On a sudden impulse he jumped on the coach and was in London that evening with no friends and only half-a-crown in his pocket.

Cobbett was always a happy man. 'No one,' he wrote later, 'has passed a happier life than I have done.' But he was also a lucky one. So now, a penniless young lad up from the country, he was approached as he sat at dinner by a fellow passenger on the coach, a hop-merchant from Southwark who, it being a small world, had had dealings with Cobbett's father. This man put the young Cobbett up, and after failing to persuade him to return home to his father found him a job in a lawyer's office in Gray's Inn.

Cobbett spent nine months working as a clerk 'perched upon a great high stool', struggling to decipher his employer's ornate handwriting. He was miserable and seeing a recruiting poster one Sunday as he was walking in St James's Park he decided to join the marines. He went to Chatham and signed on, by mistake, in a regiment of the line, then serving in Nova Scotia. It was over a year before Cobbett was posted abroad and he spent his leisure time at Chatham teaching himself English. Grammar now became one of his hobby-horses. All his life he attached the greatest importance to the writing of good English. He said later, in his *English Grammar*:

He who writes badly thinks badly. Confusedness in words can proceed from nothing but confusedness in the thoughts which give rise to them. These things may be of trifling importance when the actors move in private life: but when the happiness of millions of men is at stake, they are of importance not easily to be described.

Cobbett, industrious, intelligent and now a master of the King's English, rose quickly in the ranks. He became a corporal, then clerk to the regiment and later sergeant-major. The expression 'Carry on, Sergeant Major!' will be familiar to anyone who has served in the British Army. Used specifically on the parade ground by the officer in charge it suggests the relationship between commissioned and non-commissioned officers. The commissioned officer, owing his position to class, relies on the NCO to 'carry on', in other words to sort things out and do all the dirty work. It was in the ranks of the army that Cobbett, like many others before and since, first conceived a contempt for the officers of this world.

He called them 'the epaulet gentry', and despised them for 'their gross ignorance and vanity', 'their drunkenness and rapacity'. Cobbett's indignation was quickly aroused when he uncovered a fiddle being worked by the quartermaster, who was keeping for himself a quarter of the provisions which he was supposed to issue to the men. Like all such things it had been going on for years and everyone knew about it, but Cobbett, being the sort of man he was, decided it was his duty to expose the racket. He amassed enough evidence for his purpose, left the army when the regiment returned to England in 1791, when he was twenty-eight and registered a formal complaint with the War Office. Sensing the presence of a dangerous adversary, the Establishment reacted quickly. Cobbett's chief witness was silenced and, some other soldiers were suborned to prove that Cobbett was a traitor. He fled to France and thence after six months to America.

Later his enemies were to make much of this retreat, but Cobbett was never a coward. He had just married Anne Reid, the daughter of a Royal Artillery sergeant, and he was no doubt thinking of her welfare as much as his own when he went to France. Unlike many great radicals, Wilkes or Paine for example, Cobbett was 'a family man' who took his duties as a husband and a father very seriously. Once when his wife was pregnant and could not sleep because of the noise of dogs outside, Cobbett spent the whole night walking up and down the street in his bare feet keeping the dogs away by throwing stones at them. He helped his wife with the housework and the baby and boasted that he never once left her on her own without just cause. He hated schools and believed that a man should educate his own children. As always, he practised what he preached. Certain of his actions are only intelligible when such facts are borne in mind.

On his arrival in America in 1792 Cobbett began his long career as a writer. In a series of pamphlets he attacked the founders of the American republic and the French revolutionaries while vigorously defending England and George III. Not surprisingly he was soon up to his neck in libel actions. When he was fined $5,000 for accusing Dr Benjamin Rush of killing his patients he returned to England in 1800.

He was now thirty-seven and just getting into his stride. He began to publish a daily paper called the *Porcupine*. Cobbett was a firm supporter of the war against France and when he attacked the Peace of Amiens the London mob broke his windows and wrecked his office in Southampton Street. Shortly after, following difficulties with advertising and distribution, then a government monopoly, the *Porcupine* folded. In 1802 Cobbett began his famous *Political Register* which he was to write without a break for the rest of his life. He continued to attack the Peace and once more had his windows broken by the mob. Two years later, in 1804, Cobbett was prosecuted by the government for criminal libel, but was eventually let off. He continued to attack Pitt, in particular over corruption in the Naval Departments. This scandal led him to take an interest in nepotism generally, and the granting of pensions. Gradually Cobbett came to see that the whole of public life was riddled with corruption. His attacks became more general as he himself became more radical. Cobbett now formulated the concept of 'THE THING', by which he meant the network of MPs, stockbrokers, money-lenders, placemen and hacks bound together by mutual interests, who were running the country. The parliamentary system by which seats in the House of Commons were sold to the highest bidder was, Cobbett thought, the main source of England's troubles.

With the advent of a Tory ministry in 1807 Cobbett's diatribes grew bolder. He latched onto a new scandal when the Duke of York was accused of allowing his mistress to sell commissions. The Duke was forced to resign and the government decided that Cobbett must be silenced. An opportunity came when in 1809 he attacked the flogging by German militiamen of some mutinous soldiers at Ely. Cobbett was prosecuted for criminal libel, found guilty, fined £1,000 and sentenced to two years' imprisonment. It was a Draconian sentence but Cobbett was not downcast. During his two years in Newgate he continued to write the *Register* and to manage his Surrey farm by remote control:

> I had a farm in hand. It was necessary that I should be constantly informed of what was doing. I gave *all the orders* whether as to purchases, sales, ploughing, sowing, breeding, in short, with regard to everything, and the things were in endless number and variety, and always full of interest. My eldest son and daughter could now write well and fast. One or the other of these was always at Botley, and I had with me, having hired the best part of the keeper's house, one or two besides, either their brother or sister. We had a hamper with a lock and two keys, which came up once a week, or oftener, bringing me fruit and all sorts of country fare ... This hamper, which was always at both ends of the line looked for with the most lively interest, became our *school*. It brought me a *Journal of Labours, proceedings* and *occurrences,* written on paper of shape and size uniform, and so contrived, as to margins, as to admit of binding. The Journal used, when my eldest son was the writer, to be interspersed with drawings of our dogs, colts, or anything that he wanted me to have a correct idea of. The hamper brought me plants, bulbs,

and the like, that I might see the size of them, and almost every one sent his or her *most beautiful flowers,* the earliest violets and primroses and cowslips and bluebells, the earliest twigs of trees, and in short everything that they thought calculated to delight me.

On his release from Newgate in 1812 Cobbett kept on with his *Register.* The years that followed Waterloo were a time of great instability. The Luddites and the Parliamentary reformers were in full swing. There were renewed riots in London in 1816 and the Government in a panic suspended the Habeas Corpus Act. Cobbett who was also at this time heavily in debt, decided to take refuge in America, and in March 1817 he embarked for New York with his sons William and John. He stayed for two years and came as close to taking a holiday as he ever did. He rented a farm on Long Island. He farmed, grew vegetables and wrote his *English Grammar* of which Hazlitt said, rightly, that it was as 'entertaining as a story book'. By 1825 Cobbett could boast that it had sold 55,000 copies 'without ever having been mentioned by the old shuffling bribed sots, called Reviewers'. But Cobbett loved England. 'I myself am bound to England for life,' he wrote in his account of his year's residence in America. 'My notions of allegiance to country: my great and anxious desire to assist in the restoration of her freedom and happiness; my opinion that I possess, in some small degree, at any rate, the power to render such assistance: and, above all the other considerations, my unchangeable attachment to the people of England, and especially those who have so bravely struggled for our rights; these bind me to England.'

Cobbett returned to England in 1820, bringing with him the remains of Tom Paine, who had died in poverty and squalor in America. Determined to make amends for his previous attacks on the great Radical, Cobbett personally dug up his coffin and took the bones with him to Liverpool. His enemies were delighted by this rather ludicrous incident

and he was subjected to much ridicule in the press. Cobbett
quickly lost his enthusiasm for the bones. They were put to
one side and were last heard of, after his death, dumped in the
offices of the Official Receiver.

In 1820 Cobbett decided to stand for Parliament at Coventry.
He was bitterly opposed and lost the election. The cost of the
campaign, coupled with the effects of Lord Castlereagh's
monstrous Stamp Act of 1819 which had slapped an enormous
tax on small publications, forced Cobbett into bankruptcy.
He sold his farm at Botley and moved to Kensington where he
started a seed farm and began a period of intensive writing.
In the next ten years, he wrote his *Sermons,* the *Cottage Economy,*
the *English Gardener,* the *Woodlands,* the *French Grammar, A History
of the Protestant Reformation,* which had an enormous sale, and
the *Advice to Young Men.* In 1821 he had begun his *Rural
Rides* the account of which was to be published, as a whole,
in 1830. In 1826 he again stood for Parliament at Preston but
failed. His fortunes fluctuated and in 1830 he was once more
prosecuted for criminal libel. It was the year of the great rising
of agricultural labourers which saw rick-burning and the
smashing of machinery throughout south-east England. The
prosecution of Cobbett for fomenting the riots was a grotesque
blunder, for though he had championed the workers' cause,
he had always been at great pains to discourage acts of violence.
He subpoenaed most of the Cabinet including the Prime
Minister, Melbourne, humiliated the Chancellor Lord
Brougham and made a prolonged and crushing attack on the
Whig ministry, ending with the words:

> Whatever may be the verdict of the jury, if I am doomed
> to spend my last breath in a dungeon, I will pray God to
> bless my country: I will curse the Whigs and leave my
> revenge to my children and the labourers of England.

It was his finest hour. He was acquitted and his much-
publicised triumph over the Government brought him great

fame and even a little money. In 1832 he finally achieved his lifetime's ambition when he was elected to the Reformed parliament as member for Oldham. His maiden speech began: 'It appears that since I have been sitting here I have heard a great deal of unprofitable discussion.' He was by now nearly seventy and in the spring of 1834 fell ill. He recovered but, a year later, became again subject to attacks of severe coughing and hoarseness. He retired to his farm in Hampshire and died on 25 May. Just before his death he asked to be taken round the farm, and as his sons carried him through the fields a small boy wearing a blue smock-frock passed them by. Cobbett smiled and, his son wrote, 'seemed refreshed at the sight of the little creature which he had once precisely resembled, though now at such an immeasurable distance'.

Like all egoists Cobbett was infuriating on occasions. He barged through life, bluffing and blustering when the need arose. But his happy disposition, so much in contrast to his hectic career, endeared him to many of his contemporaries. Hazlitt wrote: 'The only time I ever saw him he seemed to me a very pleasant man – easy of access, affable, clear-headed, simple and mild in his manner, deliberate and unruffled in his speech, though some of his expressions were not very qualified. His figure is tall and portly. He has a good, sensible face – rather full, with little grey eyes, a hard square forehead, a ruddy complexion, with hair grey or powdered; and had on a scarlet broadcloth waistcoat with the flaps of the pockets hanging down, as was the custom for gentlemen-farmers in the last century.'

'One must cultivate one's garden.' Voltaire's saying has often been invoked by those who wish to avoid an involvement with public affairs. It was to Cobbett's great credit that he found time to do both: to take the problems of the nation on his shoulders and, at the same time, to cultivate his garden. He retained all his life a love of the countryside, of farming and of gardening. He wrote: 'I was brought up under a father, whose talk was chiefly about his garden and his fields, with

regard to which he was famed for his skill and his exemplary
neatness. From my very infancy, from the age of six years,
when I climbed up the side of a steep sand-rock, and there
scooped me out a plot four feet square to make me a garden,
and the soil for which I carried up in the bosom of my little
blue smock-frock (or hunting shirt), I have never lost one
particle of my passion for these healthy and rational and
heart-cheering pursuits, in which every day presents something
new, in which the spirits are never suffered to flag and in which
industry, skill and care are sure to meet with their due reward.
I have never, for any eight months together, during my whole
life, been without a garden.'

He was probably at his happiest at his farm at Botley, near
Southampton, which he bought in 1804 but which he later had
to sell to meet his debts. Miss Mitford in her *Recollections of a
Literary Life* has left an account of Cobbett at his country
home:

> He had at that time a large house at Botley, with a lawn
> and gardens sweeping down to the Bursledon River . . .
> His own house – large, high, massive, red and square, and
> perched on a considerable eminence – always struck me
> as being not unlike its proprietor . . . I never saw
> hospitality more genuine, more simple, or more thoroughly
> successful in the great end of hospitality, the putting of
> everybody completely at ease. There was not the slightest
> attempt at finery, or display, or gentility. They call it a
> farm house, and everything was in accordance with the
> largest idea of a great English yeoman of the old time.
> Everything was excellent, everything abundant – all
> served with the greatest nicety by trim waiting-damsels,
> that of the large circle of guests not one could find himself
> in the way.

The description shows that in domestic matters, as much as
in his public life, Cobbett was prepared to practise what he

preached. His ripest abuse was reserved for those whose conduct was at odds with their stated principles. He hated theorists such as economists, those who, as he said, 'have a notion that there may be great public good though producing individual misery'. Humbug of any kind Cobbett had an eye for. Once he spotted it, his abuse was merciless. He has been accused of prejudice. But if prejudice is the result of irrational subconscious emotions then Cobbett was not a prejudiced man. His 'pet hates' all had their origin in the knowledge that he had himself acquired. His fiercely held convictions were rooted in his own experience.

A hatred of theory and strong opinions based on his own findings Cobbett evinces as much in his rural writing as in his political. Almost everything he says on these topics is the result of personal trial and error. 'There is no part of the business,' he said of gardening, 'which first or last I have not performed with my own hands.' His writings therefore have a directness, regardless of their strictly horticultural merit. Autobiography keeps breaking in. The pet hates and hobby-horses are here too. Cobbett attacked tea and potatoes as vehemently as he did Wilberforce. With equal passion he championed sweetcorn and the acacia tree, both of which he introduced in large quantities into England from America. He was certain that in years to come he would be remembered chiefly for these two plants. In the event, he was wrong. But this does not invalidate all of what he wrote about them. As he says in his diatribe on potatoes, fashion plays a part in dictating what trees and vegetables are grown, and he could still be vindicated.

Cobbett lived in the Romantic Age and was himself a romantic. But he was perhaps a healthier kind of romantic than Shelley or Coleridge. He looked back to the Middle Ages and to Merry England. Before Tawney he saw clearly the connection between Religion and the Rise of Capitalism and traced the troubles of England to the Reformation. His ideal was to restore a rural property-owning democracy. His books on gardening and farming were written partly with this

aim in view, to help a man as far as possible be economically
self-sufficient.

Cobbett was engulfed by the Industrial Revolution. 'After
him radicalism is urban – and Toryism suburban,' said
G. K. Chesterton. He remains a heroic figure fighting a losing
battle against the tide of history. With his nostalgia, his love
of the countryside, his philistinism and strong views, his
sympathy with working men and hatred of parsons and
politicians he represents the stubborn spirit of the best kind of
Englishman more closely than any other character in history.

Aldworth

INTRODUCTORY OBSERVATIONS FROM A TREATISE ON COBBETT'S CORN

Men of the greatest learning, have spent their time in contriving instruments to measure the immense distance of the stars, and in finding out the dimensions and even the weight of the planets. They think it more eligible to study the art of ploughing the sea with ships, than of tilling the land with ploughs. They bestow the utmost of their skill, learnedly to pervert the natural use of all the elements, for the destruction of their own species by the bloody art of war; and some waste their whole lives in studying how to arm death with new engines of horror, and inventing an infinite variety of slaughter but think it beneath *men of learning (who only are capable of doing it), to employ their learned labours in the invention of new, or even in improving the old, means for the increasing of bread.*

Jethro Tull, *Horse Hoeing Husbandry*

INTRODUCTORY OBSERVATIONS FROM A TREATISE ON COBBETT'S CORN*

The motive for writing and publishing a work like this, if it had wanted suggestion from any mind but my own, would have been found in the *motto*, which I have chosen upon this occasion; and which is taken from him, whose work has done more to promote good agriculture, than all the other words, of all countries, put together. Certainly it is worthy of observation, and, indeed, of censure, that so very few men of learning (I mean of that sort of learning which is exhibited in, or to be acquired from books) have employed any portion of their time and talent in treating of the means of making an addition to the quantity of human food. Nay, some of them have actually prided themselves upon their ignorance of everything relating to agriculture, that first and greatest employment of man. The late Lord Erskine annually attended the sheep-shearing festivals of Mr. Coke, as long as that gentleman thought proper to treat the nobility and gentry to such festivals, and the 'learned Lord,' annually, upon those occasions, made it his boast, that he was so perfectly ignorant with regard to matters connected with the cultivation of the land, that he once, upon seeing a field of lavender, thought it was a field of wheat. To descend a step lower, Sir Walter Scott, well worthy of the first baronetcy bestowed by the present King; this

*Sweetcorn, or maize

Baronet, of book-making and book-selling notoriety, as the
affairs of Constable and Co. can bear witness; this worthy
Baronet, in a letter to Lord Somerville, published by his
Lordship in a book about oxen and sheep, took occasion,
very unnecessarily, and even ostentatiously, to observe, that,
as to the business of agriculture, he hardly knew more than the
pen with which he was writing. To descend lower still ('where
will you stop, then?' the reader will exclaim), Mr. Adolphus of
the George-the-Third-history-fame, or rather obscurity, in
pleading against me for one Farlar, before the Secondary of
London, took occasion, while he displayed his white handker-
chief, and his genteely-pale hand, to tell the Jury, that, as to
brewing, or the sort of materials made use of in that process;
or, as to any rural affair, of whatever description, he himself
had no knowledge; and that, in fact, if taken into the fields
or the gardens, he knew not any one plant from any other;
which declaration if made before people in the country, before
a set of farmers' men and maids, would convince them that the
orator was fit for little else besides being knocked on the head.

When learned gentlemen, and learned lords, and girl-
bewitching novel-writers, make this kind of ignorance their
boast, is it any wonder that it should become the fashion,
with persons in the middle rank of life, to make it their boast,
that they do not know what you mean, when you talk of such
things as spades and ploughs, and rakes and harrows? It is
curious, too, that these same persons are ashamed to be
thought ignorant of the histories of all the nations on earth;
and, as to *politics*, they all understand *politics;* they would be
ashamed not to be thought clearly to understand that which
the King and the Parliament ought to do. The constitution! Oh!
they all understand that, though made up of a series of maxims,
decisions, and positive acts, grown together in the course of
twelve hundred years. And, *religion,* now! what man of them
would not be ashamed, not to be thought competent to decide,
not only between Calvin and the Pope; but to determine, to a
hair's breadth, the right and wrong of all the intervening

classes, amounting to about fifty in number, each differing in its creed from all the other forty-nine?

Mr. Tull, in the elegant passage which I have taken for my motto, observes on the waste of learning, in measuring and weighing the stars; which brings to my recollection, that, when I was in the army, in New Brunswick, I was acquainted with a serjeant, who was a young man, and who, as well as myself, was a great reader; but he was smitten with astronomy, and wanted me to pay attention to some of the discoveries he had made. 'I tell you what,' said I, 'I do not care what they are doing up there; their orders, whether general, garrison, or regimental, can never affect me: study you, if you please, what they are about, I will confine my studies to things which pass upon the earth'. In our schools, nay, in our universities, every thing is taught, but that which is the most useful and honourable of all; namely, the means of raising food, drink, and clothing, and materials for buildings; without which, mankind must cease to exist; and without the whole of which in tolerable perfection and abundance, no nation can be either great or happy. We have lectures upon everything but agriculture: Doctor Birkbeck treats us to a theory of the winds; another takes infinite pains to explain how the air is pumped out of the body of a rat; there are a great number of lecturers to teach us how the veins and the intestines are formed, and how they ought to be emptied; but not a soul to tell us the best way of filling them. Commissioners of Scotch herrings we have had, and a Count of the White-Eagle (Rumford), together with a whole troop of soup-kettle philosophers, to teach us how to stew old bones into jelly, and by how little the human body can be sustained; but not a single man to give us a lecture on the means of providing that plenty, that abundance of good living, without which man had better be dead than alive, and for which our country was, for so many ages, so famous.

The great study, of late years, appears to have been to discover the means of reducing the most numerous and most useful class of the people to exist upon the smallest possible quantity of food; and, failing here, Parson Malthus has suggested the means, improved upon by the infamous Peter Thimble, and the equally infamous Carlile, of checking the course of nature in the producing of children. The Parson and his worthy coadjutors never seem to have thought, for a single moment, of a more just distribution of the food already raised, and still less of any means of adding to the quantity. The schemes of these worthies not being attended with success, schemes for *getting rid of the people,* by sending them out of the country, have, at last, been resorted to, and have actually been brought before Parliament, by Mr. Wilmot Horton, the patron of the project.

My efforts have, all my life long, since I became a man, been directly the reverse of those of these projectors. I have used various endeavours to cause an addition to be made to the food, the drink, the raiment, of the industrious classes. I know, and I have always known, that complete success cannot attend these endeavours, so long as the present, or any thing like the present, burthen of taxes remains; but, a change in this respect must come; and, in the meanwhile, it is my duty to persevere in my efforts to add to the permanent resources, the permanent strength, and the permanent happiness of my country.

GARDENING

GARDENING

THE SITUATION

If one could have what one wished, in point of situation, from the wall on the north side of the garden, after a little flat of about a rod wide, one would have a gentle slope towards the south, about thirty feet in width. The remainder of the ground, to the wall on the south side of the garden, one would have on a true level. The gentle slope contributes to early production; and though it is attended with the inconvenience of washing, from heavy rains, that inconvenience is much more than made up for by the advantage attending the circumstances of earliness. I recollect the ancient kitchen-garden which had been that of the monks, at Waverley Abbey. It lay full to the south, of course; it had a high hill to the back of it, and that hill covered with pretty lofty trees. The wall on this north side of the garden was from twelve to fourteen feet high, built partly of flints, and partly of the sand-stone, which is found in abundance in the neighbourhood, and it was about three feet through, even at the top. The ground of which the garden consisted had been the sloping foot of a hill, taking in a part of the meadow that came after the hill, and lay between it and the river Wey. A flat of about twenty feet wide had been made on the side of the hill and, at the back of this flat, the wall was erected. After the flat, towards the south, began the slope; at the end of the slope began the level ground, which grew more and more moist as it approached the river. At the foot of the garden there ran a rivulet, coming from a fish-pond, and at a little distance from that, emptying itself into the river. The hill itself was a bed of sand; therefore, the flat, at the back of which the north wall stood; that is to say, the wall on the north side of the garden; this flat must have been *made* ground. The slope must have been partly made, otherwise it would have been too sandy.

This was the finest situation for a kitchen-garden that I ever saw. It was wholly torn to pieces about fifty years ago; the wall pulled down; the garden made into a sort of lawn, and the lower part of it, when I saw the spot about three years ago, a coarse, rushy meadow, all the drains which formerly took away the oozings from the hill having been choked up or broken up; and that spot where the earliest birds used to sing, and where prodigious quantities of the finest fruits used to be borne, was become just as sterile and as ill-looking a piece of ground, short of a mere common or neglected field, as I ever set my eyes on. That very spot where I had seen bushels of hautboy strawberries, such as I have never seen from that day to this; that very spot, the precise locality of which it took me (so disfigured was the place!) the better part of an hour to ascertain, was actually part of a sort of swampy meadow, producing sedgy grass and rushes. The most secluded and beautiful spot was given away by the ruthless tyrant, Henry the Eighth, to one of the basest and greediest of his cormorant courtiers, Sir William Fitzwilliams; it became afterwards, according to Grose, the property of the family of Orby Hunter; from that family it passed into the hands of a Sir Robert Rich, much about fifty years ago. The monastery had been founded by Giffard, bishop of Winchester, who brought to inhabit it the first community of Cistercian monks that were settled in England. He endowed the convent at his own expense; gave it the manor and estate, and gave it also the great tithes of the parish of Farnham, in which it lies. A lofty sand-hill sheltered it to the north; others, in the form of a crescent, sheltered it to the east. It was well sheltered to the west; open only to the south, and a little to the south-west. A valley let in the river

Wey at one end of this secluded spot, and let it out at the other
end. Close under the high hill on the north side, a good
mansion-house had been built by the proprietors who succeeded the monks, and these proprietors, though they had
embellished the place with serpentine walks and shrubberies,
had had the good taste to leave the ancient gardens, the grange,
and as much of the old walls of the convent as was standing;
and, upon the whole, it was one of the most beautiful and
interesting spots in the world. Sir Robert Rich tore everything
to atoms, except the remaining wall of the convent itself. He
even removed the high hill at the back of the valley itself;
actually carried it away in carts and wheelbarrows; built up a
new-fashioned mansion-house with grey bricks, made the place
look as bare as possible; and, in defiance of nature, and of all
the hoar of antiquity, made it very little better than the vulgar
box of a cockney.

I must be excused for breaking out into these complaints. It
was the spot where I first began to learn to work, or rather,
where I first began to eat fine fruit in a garden; and though I
have now seen and observed upon as many fine gardens as any
man in England, I have never seen a garden equal to that of
Waverley. Ten families, large as they might be, including
troops of servants (who are no churls in this way) could not
have consumed the fruit produced in that garden. The peaches,
nectarines, apricots, fine plums, never failed: and, if the workmen had not lent a hand, a fourth part of the produce never
could have been got rid of. Sir Robert Rich built another
kitchen garden, and did not spare expense; but he stuck the
walls up in a field, unsheltered by hills and trees; and though
it was twice the size of the monks' garden, I dare say it has
never yielded a tenth part of the produce.

It is not everywhere that spots like this are to be found; and
we must take the best that we can get, never forgetting, however,
that it is most miserable taste to seek to poke away the kitchen-
garden, in order to get it out of sight. If well managed, nothing
is more beautiful than the kitchen-garden: the earliest blossoms

come there; we shall in vain seek for flowering shrubs in March, and early in April, to equal the peaches, nectarines, apricots and plums; late in April, we shall find nothing to equal the pear and the cherry; and, in May, the dwarf, or espalier, appletrees, are just so many immense garlands of carnations. The walks are unshaded: they are not greasy or covered with moss, in the spring of the year, like those in the shrubberies: to watch the progress of the crops is by no means unentertaining to any rational creature; and the kitchen-garden gives you all this long before the ornamental part of the garden affords you anything worth looking at. Therefore, I see no reason for placing the kitchen-garden in some out-of-the-way place, at a distance from the mansion-house, as if it were a mere necessary evil, and unworthy of being viewed by the owner. In the time of fruiting, where shall we find anything much more beautiful to behold than a tree loaded with cherries, peaches, or apricots, but particularly the two latter? It is curious enough that people decorate their chimney-pieces with imitations of these beautiful fruits, while they seem to think nothing at all of the originals hanging upon the tree, with all the elegant accompaniments of flourishing branches, buds and leaves.

We must take, as I said before, the best ground that we have, and, for my part, I would take it almost anywhere except in the front of a mansion-house. It must absolutely be open to the south: well sheltered, if it can be, from the north and from the east; but open to the south it must be, or you can have neither fine wall fruit, nor early crops of garden-plants. If you can have the slope such as I have described it to have been at Waverley, it is easy to make a flat before the face of the wall, on the north side of the garden: but, to have the *whole* of a garden upon a slope is by no means desirable; for, however gentle the slope may be, the water will run off; and, in certain cases, it is absolutely necessary that the water should not run away; but have time to soak gently into the ground. I have had great opportunity of acquiring knowledge in this respect. Part of my ground at Kensington forms a very gentle slope. The soil of this slope is as good, both at top and bottom, as any ground in the world; but I have always perceived that seeds never rise there with the same alacrity and the same vigour as they do upon the level part, though there the soil is much inferior. That is particularly the case with regard to strawberries, which will grow, blow like a garland and even bear pretty numerously on the side of a bank where scarcely any moisture can lodge; but which I have never seen produce large and fine fruit except upon the level. The same may be said of almost every garden plant and tree; and, therefore, if I could avoid it, I would always have some part of a garden not upon the slope. Slopes are excellent for early broccoli, early cabbages, winter spinage, onions to stand the winter, artichokes to come early, early peas, early beans and various other things; but there ought to be some part of the garden upon a true level; for, when the month of June comes, that is the part of the garden which will be flourishing.

As to shelter, hills, buildings, lofty trees, all serve for the purpose; but the lofty trees ought not to stand too near. They ought not to shade by any means; and none of their leaves ought to drop into the garden. Leaves from such trees, blown into the garden by high winds, are merely a temporary inconvenience; but shade would do injury, though, perhaps, if not too deep, counterbalanced by the warmth and the shelter that the trees would afford.

Before I quit this subject of *situation* I cannot refrain from attempting to describe one kitchen-garden in England, to behold which is well worth the trouble and expense of a long journey, to any person who has a taste in this way: I mean that of Mr. Henry Drummond at *Albury*, in the county of Surrey. This garden is, in my opinion, nearly perfection, as far as relates to situation and form. It is an oblong square; the wall on the north side is close under a hill; that hill is crowned with trees which do not shade the garden. There is a flat, or terrace, in the front of this wall. This terrace consists, first of a border for the fruit trees to grow in, next of a broad and beautiful gravel walk, then, if I recollect rightly, of a strip of short grass. About the middle of the length, there is a large basin supplied with water from a spring coming out of the hill, and always kept full. The terrace is supported, on the south side of it, by a wall that rises no higher than the top of the earth of the terrace. Then comes another flat, running all the way along; this flat is a broad walk, shaded completely by two rows of yew-trees, the boughs of which form an arch over it so that here, in this kitchen-garden, there are walks for summer as well as for winter: on the gravel walk you are in the sun, sheltered from every wind; and, in the yew-tree walk you are completely shaded from the sun in the hottest day in summer. From the yew-tree walk the ground slopes gently down towards the brook which runs from Sheer through Albury, down to Chil-worth; where, after supplying the paper-mills and powder-mills, it falls into the river Wey. The two end walls of the garden have plantations of trees at the back of them; so that, except that here is no ground, except the terrace, which is not upon the slope, this garden, which is said to have been laid out by Sir Philip Evelyn for some member of the family of Howard, is everything that one could wish. The mansion-house stands at a little distance opposite the garden, on the other side of the brook; and, though all the grounds round about are very pretty, this kitchen-garden constitutes the great beauty of the place. Here, too, though Evelyn *might* have *revived*, this charming spot was chosen, the garden was *made*, and the cloister of yew-trees planted, by the *monks* of the Priory of St Austin, founded here in the reign of Richard I., and the estates of which Priory were given by the bloody tyrant to Sir Anthony Brown.

BOX EDGING

It is impossible to dig the ground close to a walk which has not a sufficient protection, without bringing dirt upon the walk: all the shovelling in the world will not get it off again clean, unless you go down so deep as to take up part of the gravel with the dirt; so that, your walk must soon become a dirty-looking

affair, in which weeds and grass will be everlastingly coming:
or you must take away, little by little, the gravel, by shovelling,
till you have flung it pretty nearly all upon the borders and
flats, and thereby not only destroyed your walk, but injured
your cultivated land. To prevent these very great troubles and
injuries, you must resolve to have an efficient protection for the
walk; and this, I venture to assert, is to be obtained by no
other means than by the use of Box. Many contrivances have
been resorted to for the purpose of avoiding this pretty little
tree, which, like all other really valuable things, requires some
little time; some little patience, and great attention, after you
have got it. In the end, indeed, it is a great deal cheaper than
anything else; but it requires some attention and patience at
first, and regular clipping every year twice. I have seen, and
have had, as an *edging* (which ramparts of this sort are called),
a little flowering plant called thrift: I have seen strawberries
thickly planted for this purpose: I have seen daisies, and various
other things, made use of as edgings: but, all these herbaceous
things ramble very quickly over the ground; extend their creepers
over the walk, as well as over the adjoining ground; and,
instead of being content to occupy the space of three inches
wide, to which it is vainly hoped their moderaton will confine
them, they encroach to the extent of a foot the first summer;
and, if left alone for only a couple of years, they will cover the
whole of a walk six feet wide, harbouring all sorts of reptiles,
making the walk pretty nearly as dirty as if it did not consist of
gravel. I have sometimes seen narrow edgings of grass, which,
perhaps, are the worst of all. Make such an edging, of four
inches wide, in the autumn, and it will be sixteen inches wide
before the next autumn, unless you pare down the edges of it
three or four times. This must be done by a line; and even then,
some dirt must be cut from the edging, to come into the walk:
this is, in fact, a rampart of dirt itself. It must be mowed not
less than ten times during the summer, or it is ugly beyond
description; besides bringing you an abundant crop of seeds to
be scattered over the walk, and over the adjoining ground. Of

all edgings, therefore, this is the least efficient for the purpose, and by far the most expensive.

The box is at once the most efficient of all possible things, and the prettiest plant that can possibly be conceived: the colour of its leaf; the form of its leaf; its docility as to height, width, and shape; the compactness of its little branches; its great durability as a plant; its thriving in all sorts of soils, and in all sorts of aspects; its freshness under the hottest sun, and its defiance of all shade and all drip: these are beauties and qualities, which, for ages upon ages, have marked it out as the chosen plant for this very important purpose.

The box, to all its other excellent qualities, adds that of facility of propagation. You take up the plants, when they are from three to six inches high, when they have great numbers of shoots coming from the same stem; you strip these shoots off, put them into the ground, to about the depth of two inches, or a little more; fasten them well there, first with the hand, and then with the foot; clip them along at the top to within about two inches of the ground, and you have a box edging at once. You must, indeed, purchase the plants, if you have not taken care to raise them before-hand; and, as to thrift, strawberries, daisies, or grass edgings, there are generally cart-loads of them to be thrown away, or to be dug from a common. I should suppose, however, that ten pounds' worth of box, bought at the nurseries, would be sufficient for the whole garden; and, then, with common care, you have neat and efficient edgings for a life-time.

To plant the box, some care must be taken. The edging ought to be planted as soon as the gravel walks are formed. The box ought to be placed perpendicularly, and in a very straight line, close to the gravel; and with no earth at all between it and the gravel. It ought to stand when planted and cut off, about four inches high; and the earth in the borders or plats ought to be pushed back a little, and kept back for the first year, to prevent it from being washed back over the walks. When the edging arrives at its proper height, it will stand about seven inches high, on the gravel side, and will be about three inches higher than the earth in the border, and will act like a little wall to keep the earth out of the walks; which, to say nothing of the difference in the look, it will do, as effectually as brick, or boards, or any thing else, however solid. The edging ought to be clipped in the winter, or very early in the spring, on both the sides and at top; a line ought to be used to regulate the movements of the shears; it ought to be clipped again, in the same manner, just about Midsummer; and, if there be a more neat and beautiful thing than this in the world, all that I can say, is, that I never saw that thing.

THE GREEN-HOUSE

As to the making of green-houses, I shall think of nothing more than a place to preserve tender plants from the frost in the winter, and to have hardy flowers during a season of the year when there are no flowers abroad. It is necessary, in order to make a green-house an agreeable thing, that it should be very near to the dwelling-house. It is intended for the pleasure, for the rational amusement and occupation, of persons who would otherwise be employed in things irrational; if not in things mischievous. To have it at a distance from the house would be to render it nearly useless; for, to take a pretty long tramp in the dirt or wet, or snow, to get at a sight of the plants, would be, nine times out of ten, not performed; and the pain would, in most instances, exceed the pleasure. A green-house should, therefore, be erected against the dwelling-house. The south side of the house would be the best for the green-house; but any aspect, to the south of due east and due west may do tolerably well; and a door into it, and a window, or windows looking into it, from any room of the house in which people frequently sit, makes the thing extremely beautiful and agreeable. It must be glass on the top, at the end most distant from the house, and in the front from about three feet high. There should be an outer door for the ingress and egress of the gardener, and a little flue running round for the purpose of

obtaining heat sufficient for the keeping of a heat to between
forty and fifty degrees of Fahrenheit's thermometer. Stages,
shelves, and other things necessary for arranging the plants
upon, would be erected according to the taste of the owner,
and the purposes in view. Besides the plants usually kept in
green-houses, such as geraniums, heaths and the like, I should
choose to have bulbous-rooted plants of various sorts, even the
most common, not excluding snow-drops and crocuses.
Primroses and violets (the common single sorts, for the others
have no smell), cowslips and daisies; some dwarf roses; and
thus a very beautiful flower-garden would be seen in the month
of February, or still more early. Green-house plants are always
set out of doors in the summer, when they are generally very
much eclipsed in beauty by plants of a hardy and more
vigorous description. If there be no green-house, these plants
are taken into the house, shut up in a small space, very frequently in the shade, and always from strong light, especially
early in the morning; which greatly injures, and, sometimes
totally destroys, them; besides, they really give no pleasure,
except in winter; for, as was observed before, after the month
of May comes, they are far surpassed in beauty by the shrubberies and the parterre.

Nor is such a place without its real use, for few persons will
deny that fruit is of use; none will deny that fine grapes are
amongst the best of fruit; we all know that these are not to be
had in England, in the general run of years, without the
assistance of glass, and the green-house, in which the shade of
the grapes would do no injury to the plants, because these
would be out in the open air, except at the time when there
would be little of leaf upon the vines, is as complete a thing
for a grapery as if made for that sole purpose; for, if the heat
of from forty to fifty degrees would bring the vines to bear at a
time, or, rather, to send out their leaves at a time inconvenient
for the plants, you have nothing to do but to take the vine
branches out of the house, and keep them there until such time
that they might be put in again without their leaves producing

an inconvenient shade over the plants, previous to the time of these latter being moved out into the open air.

As the green-house would have given you a beautiful flower-garden and shrubbery during the winter, making the part of the house to which it is attached the pleasantest place in the world, so, in summer, what can be imagined more beautiful than bunches of grapes hanging down, surrounded by elegant leaves, and proceeding on each grape from the size of a pin's head to the size of a plum? . . . I cannot conclude without observing, that it is the *moral* effects naturally attending a green-house, that I set the most value upon. I will not, with Lord Bacon, praise pursuits like these, because 'God Almighty first planted a garden;' nor with Cowley, because 'a Garden is like Heaven;' nor with Addison, because a 'Garden was the habitation of our first parents before their fall;' all of which is rather far-fetched, and puts one in mind of the dispute between the gardeners and the tailors, as to the antiquity of their re-spective callings; the former contending that the planting of the garden took place before the sewing of the fig-leaves together; and the latter contending, that there was no garden-ing at all till Adam was expelled, and compelled to work; but, that the sewing was a real and bona fide act of tailoring. This, to be sure, is vulgar and grovelling work; but who can blame such persons when they have Lord Bacon to furnish them with a precedent? I like, a great deal better than these writers, Sir William Temple, who, while he was a man of the soundest judgment, employed in some of the greatest concerns of his country, so ardently and yet so rationally and unaffectedly praises the pursuits of gardening, in which he delighted from his youth to his old age; and of his taste in which he gave such delightful proofs in those gardens and grounds at Moor Park in Surrey, beneath the turf of one spot of which he caused, by his will, his heart to be buried, and which spot, together with all the rest of the beautiful arrangement, has been torn about and disfigured within the last fifty years by a succession of wine-merchants, spirit merchants, West Indians, and God

knows what besides: I like a great deal better the sentiments of
this really wise and excellent man; but I look still further as to
effects. There must be amusements in every family. Children
observe and follow their parents in almost everything. How
much better, during a long and dreary winter, for daughters,
and even sons, to assist, or attend, their mother, in a green-
house, than to be seated with her at cards, or in the blubber-
ings over a stupid novel, or at any other amusement that can
possibly be conceived! How much more innocent, more pleasant, more free from temptation to evil, this amusement, than
any other! How much more instructive too! 'Bend the twig
when young:' but, here, there needs no force; nay, not even
persuasion. The thing is so pleasant in itself; it so naturally
meets the wishes; that the taste is fixed at once, and it remains,
to the exclusion of cards, and dice, to the end of life. Indeed,
gardening in general is favourable to the well-being of man.
As the taste for it decreases in any country, vicious amusements and vicious habits are sure to increase. Towns are preferred to the country; and the time is spent in something or
other that conduces to vice and misery. Gardening is a source
of much greater profit than is generally imagined; but, merely
as an amusement, or recreation, it is a thing of very great value:
it is a pursuit not only compatible with, but favourable to, the
study of any art of science: it is conducive to health, by means
of the irresistible temptation which it offers to early rising; to
the stirring abroad upon one's legs; for a man may really ride
till he cannot walk, sit till he cannot stand, and lie abed till he
cannot get up. It tends to turn the minds of youth from amusements and attachments of a frivolous or vicious nature: it is a
taste which is indulged at home: it tends to make home pleasant,
and to endear us to the spot on which it is our lot to live: and,
as to the *expenses* attending it, what are all these expenses,
compared with those of the short, the unsatisfactory, the
injurious enjoyments of the card-table, and the rest of
those amusements or pastimes which are sought for in the
town?

SOME HERBS AND VEGETABLES

Basil, is a very sweet annual pot-herb, being of two sorts, the dwarf and the tall. It should be sowed in very fine earth early in the spring and transplanted into earth equally fine, with very great care. But let me here speak of the place for herbs in general. They should all be collected together in one spot if possible. The best form is a long bed, with an alley on each side of it, the bed too narrow to need trampling in order to reach the middle of it. The herbs should stand in rows made across this bed, the quantity of each being in due proportion to the consumption of the family; for it is a mark of great want of judgment to occupy great spaces of ground with things that can be of no possible use. We have often seen, in a gentleman's garden, as much parsley growing as would be sufficient for the supply of a large country town; and, as to *mint*, I have often seen it covering several rods of ground, when the sensible original intention was that it should be confined within the space of a couple of square yards. Mint, however, forms an exception to what has just been said about collecting the herbs together in one place; for its encroachments are such that it must be banished to some spot where those encroachments can occasionally be restricted by the operation of the spade.

Camomile is a perennial medicinal herb of great use. It may be propagated from seed, but it is most easily propagated by parting the roots. One little bit of root will soon make a bed sufficient for a garden. The flowers which are used in medicine should be gathered before they begin to fade, and at a time when they are perfectly dry; and then put into a shady and airy place to dry, which they will do perfectly, but not in less than a month. When perfectly dry, they should be put into a paper bag, hung up in a dry place, and kept from all dust.

Chervil. This, like celery, spinage, and some few other garden plants, is very much liked by some people, and cannot be endured by others. It is an annual plant: its leaves a good deal like those of double parsley: it is used in salads, to which it gives an odour that some people like very much: it bears a seed resembling that of a wild oat; it is sowed in rows late in March or early April; and a very small patch of it is enough for any garden: it bears its seed, of course, the first summer, bears it too in great abundance, and if properly preserved, the seed will last for six or seven years at the least.

Cives. A little sort of *Onion,* which is perennial: it may be propagated from seed; but the easiest way is by parting the roots, which are bunches of little bulbs like those of crocuses or snow-drops. The green only of this plant is used; and a very small patch is sufficient for any garden. Five or six clumps in the herb-bed would be sufficient.

Fennel is a perennial herb, propagated from seed or from offsets, sowed in the spring, or the offsets planted in the fall. The plants should stand about a foot asunder. The leaves are used in salads, or for the making a part of the sauce for fish. In winter, the seeds are bruised, to put into fish-sauce, and they give it the same flavour as the leaves of the plant. It is a very hardy thing; two yards square in the herb-bed will be enough for any family; and, once in the ground, it will stand for an age.

Jerusalem Artichoke. This plant bears at the root, like a potato, which, to the great misfortune of many of the human race, is everywhere but too well known. But this artichoke, which is also dug up and cooked like a potato, has, at any rate, the merit of giving no trouble either in the cultivation or the propagation. A handful of the bits of its fruit, or even of its roots, flung about a piece of ground of any sort, will keep bearing for ever in spite of grass and weeds; the difficulty being, not to get it to grow, but to get the ground free from it when once it has taken to growing. It is a very poor, insipid vegetable; but if you have a relish for it pray keep it out of the garden, and dig up the corner of some field, or of some worth-less meadow, and throw some roots into it.

Mint. There are two sorts: one is of a darker green than the other; the former is called *pepper-mint* and is generally used for *distilling* to make mint water: the latter, which is called *spear-mint,* is used for the table in many ways. The French snip a little into their *salads*; we boil a bunch amongst green peas, to which it gives a pleasant flavour; chopped up small, and put along with sugar into vinegar, we use it as a sauce for *roasted lamb*: and a very pleasant sauce it is. Mint *may* be propagated from seed: but a few bits of its roots will spread into a bed in a year. To have it in winter, preserve it precisely like *marjoram* and instead of *chopping* it for sauce, crumble it between your fingers.

Mustard. Why buy this, when you can grow it in your garden? The stuff you buy is half drugs, and is injurious to health. A yard square of ground, sown with common Mustard, a crop of which you would grind for use, in a little mustard-mill, as you wanted it, would save you some money, and probably save your life. Your mustard would look brown instead of yellow; but the former colour is as good as the latter; and, as to the taste, the real mustard has certainly a much better one than that of the drugs and flour which go under the name of mustard. Let any one try it, and I am sure he will never use the drugs again. The drugs, if you take them freely, leave a burning at the pit of your stomach, which the real mustard does not.

Parsley. Known to every human being to bear its seed the second year, and, after that, to die away. It may be sowed at *any season* when the frost is out of the ground. The best way is to sow it in spring, and in very *clean* ground; because the seed lies long in the ground, and, if the ground be foul, the weeds choke the plants at their coming up. A bed of six feet long and four wide, the seeds sowed in drills at eight inches apart, is enough for any family in the world. This would be enough about parsley; but people want it all the year round. There are some winters that will destroy it completely if it be wholly unprotected, and there are no means of preserving it dry in the manner which has been directed for other herbs. Therefore, if you perceive sharp weather approaching, lay some peas-haulm or straw, not very thickly, over the bed, and do not take it off until after the thaw has completely taken place. The

rotting of vegetables is occasioned by thawing in the light
more than by the frost. When the thaw has completely taken
place, the peas-haulm or the straw may be taken away, and,
by these means, parsley may be safely kept through any winter,
that we have in England: for it can be thus kept even in
America where the frost goes down into the ground full
four feet.

Pea. This is one of those vegetables which all people like. From
the greatest to the smallest of gardens, we always find peas, not
to mention the thousands of acres which are grown in fields for
the purpose of being eaten by the gardenless people of the
towns. Where gardening is carried on upon a royal, or almost
royal scale, peas are raised by means of artificial heat, in order
to have them here at the same time that they have them in
Portugal, which is in the months of December and January.
Beneath this royal state, however, the next thing is to have
them in the natural ground as early as possible; and that may
be sometimes by the middle of May, and hardly ever later than
about the first week of June. The late King, George the Third,
reigned so long, that his birth day formed a sort of season with
gardeners; and, ever since I became a man, I can recollect that
it was always deemed rather a sign of bad gardening if there
were not green peas in the garden fit to gather on the fourth of
June. It is curious, that green peas are to be had as early in
Long Island, and in the sea-board part of the state of New
Jersey, as in England, though not sowed there, observe, until
very late in April, while ours, to be very early, must be sowed
in the month of December or January. It is still more curious,
that, such is the effect of habit and tradition, that, even when I
was last in America (1819), people talked just as in England
about having green peas on the King's birth-day, and were just
as ambitious for accomplishing the object; and I remember a
gentleman who had been a republican officer during the
Revolutionary War, who told me that he always got in his
garden green peas fit to eat on old Uncle George's birth-day.
This, however, is the general reason for the coming in of green

peas in England; but, to have them at this season, the very
earliest sort must be sowed; they must be sowed, too, in
November, or as soon after as the weather will permit, and
they must be sowed on the south side of a wall, or of a very
close and warm hedge, the ground not being wet in its nature
by any means. The frosts will be very apt to cut them off, and,
if the weather be mild, they will be apt to get so forward as to
be cut off in January or February. They should, therefore, be
kept earthed up a little on both sides; and, if hard frosts
approach, they should be covered with peas-haulm or straw,
and these should be taken off as soon as the thaw has completely taken place. . . . A second sowing should take place a
month or six weeks after the first. Sow again early in March,
and then once in a month or three weeks, until the end of May.
Too many should not be sowed at a time, and less of the tall
sorts than of the low sorts. The manner of sowing peas is the
same in all cases. You can make a drill with a hoe, three inches
deep, in ground as rich as you can make it, sow the peas along
not too thick, put back upon them the earth that came out of
the drill, and tread it down with your feet pretty nearly as hard
as you can, and then, especially in winter-time, keep a sharp
look-out after the mice. When the peas come up, you ought, in
all cases, to hoe the ground nicely about them, and draw a little
earth to them even immediately, drawing up more and more
earth on each side as the plants advance in height, until you
have, at last, a little rise, the top of which would be six or seven
inches above the level of the ground; this not only keeps them
upright, but supplies them with food for roots that will shoot
out of the stems of the plants. Peas must have sticks, and these
sticks must be proportioned to the height which the sorts
respectively generally attain. For the early-frame pea, two feet
and a half, or three feet, above the ground, is sufficient; for the
next in height, four or five feet. For the tall sorts, from six to
eight, and even nine feet. The distances at which the rows are
to be sowed, must be somewhat in proportion to these heights,
the smaller peas may stand at three feet apart, but the taller

ones, and especially the tallest ones of all, ought to be at six or seven feet apart at the least. You get nothing by crowding them, nor do you get anything by sowing double instead of single rows of peas. If you try it, you will find that a single plant standing out away from all others, will produce more fruit than any six plants standing in a common single row, though the soil be the same, and though the stick be of the same height. This is enough to convince any one of the mischievous effects of crowding. If you plant the taller peas at distances too close, or indeed any peas, the rows shade one another; there will be no fruit except just at the top, that part of the plant which should bear early will not bear at all, those that come at the top will be pods only about half full; and if you plant tall peas so close, and with sticks so short as to cause the wet to bend the heads of the plants down, you will literally have no fruit at all, a thing which I have seen take place a hundred times in my life-time. My gardener had once sowed, while I was from home, a piece of garden with the tall marrowfat pea, and had put the rows at about three feet apart. I saw them just after they came up. The ground was such as was very good, and which I knew would send the peas up very high; I told him to take his hoe and cut up every other row, but they looked so fine and he was so obstinate, that I let them remain, and made him sow some more at seven feet apart very near to the same place, telling him that there never could be a pea there, and that if it so turned out, never to attempt to have his own way again. Both the patches of peas were sticked in due time, and both grew very fine and lofty; but his patch began to get together at the top, and just about the time that the pods were an inch long, there came a heavy rain, smashed the whole of them down into one mass, and there never was a single pea gathered from the patch, while the other patch, the single rows of which were seven feet apart, produced an uncommonly fine and lasting crop. The destroyed patch of peas was however of previous advantage; for it made me the master of my gardener, a thing that happens to very few owners of gardens.

A sufficient distance is one of the greatest things in the raising of peas, whether they be sticked or whether they be not; and they never ought to be sowed too thickly in the row. I never tried it, but I verily believe that a row of peas, each plant being at two or three inches distance from the other, would bear a greater crop than if sowed in the usual way. At any rate, never sow too thick, on any account, at any time of the year. As to *sorts* of peas, the earliest is the *early-frame*, then comes the *early-charlton*, then the *blue-prussian* and the *hotspur*, then the *dwarf-marrowfat*, then the *tall-marrowfat*, then *knight's pea*. There are several others, but here are quite enough for any garden in the world. If all these tall sorts be sowed in March, and some more of them again in April, not too many at a time, they will come in one after another, and will keep up a regular succession until about the latter end of July, or even later. After this all peas become mildewed, and their fruit good for very little. As to *saving the seed* of peas, it is impossible to do it well in a kitchen-garden, where you must always have more than one sort of pea in bloom at the same time. If you be very curious about this matter, you must sow somewhere in the corner of a field, and not gather any of the peas to eat; but let them all stand to ripen. When ripe, they are to be threshed out and put by in a dry place. Peas want no watering, but there should be a good digging between the rows just about the time that the bloom begins to appear, for that furnishes new food to the roots at the time when it is most wanted. Great care must be taken to keep slugs and snails away from peas; for if they get amongst them and are let alone for a very little while, they bite the whole off, and they never sprout again to any good purpose.

Purslane. A mischievous weed, eaten by Frenchmen and pigs when they can get nothing else. Both use it in salad, that is to say, raw.

Potatoes. I have made no experiments as to this root, and I am now about to offer my opinions as to the mode of cultivating it. But, so much has been said and written *against me* on account of my scouting the idea of this root being proper as *food for men*, I will, out of respect for public opinion, here state my *reasons* for thinking that the Potatoe is a root *worse than useless*.

When I published some articles upon this subject, in England, I was attacked by the *Irish* writers with as much fury as the Newfoundlanders attack people who speak against the Pope: and with a great deal less reason; for, to attack a system which teaches people to fill their bellies with fish for the good of their souls might appear to be dictated by malice against the sellers of the fish; whereas my attack upon Potatoes was no attack upon the sons of St. Patrick, to whom, on the contrary, I wished a better sort of diet to be afforded. Nevertheless, I was told, in the Irish papers, not that I was a *fool*: that might have been *rational*; but, when I was, by these zealous Hibernians, called a *liar*, a *slanderer*, a *viper*, and was reminded of all my *political sins*, I could not help thinking that, to use an Irish peeress's expression with regard to her Lord, there was a little of the Potatoe *sprouting out of their head*.

These rude attacks upon me even were all *nameless*, however; and, with nameless adversaries I do not like to join battle. Of one thing I am very glad; and that is, that the Irish *do not like* to live upon what their accomplished countryman Doctor Drennan calls "Ireland's *lazy root*". There is more sound political philosophy in that poem than in all the enormous piles of Plowden and Musgrave. When I called it a *lazy root*; when I satyrized the use of it; the Irish seemed to think that their national *honour* was touched. But I am happy to find that it is not *taste*, but *necessity*, which makes them mess-mates with the pig; for when they come to this country, they invariably prefer to their *favourite root*, not only fowles, geese, ducks and turkeys, but even the flesh of oxen, pigs and sheep!

In 1815, I wrote an article, which I will here insert because it contains my opinions upon this subject. And when I have done that I will add some calculations as to the comparative value of an acre of wheat and an acre of potatoes. The article was a letter to the *Editor of the Agricultural Magazine*; and was in the following words:

TO THE EDITOR OF THE
AGRICULTURAL MAGAZINE

Sir,

In an article of your Magazine for the month of September last, on the subject of my Letters to Lord Sheffield, an article with which, upon the whole, I have reason to be very proud, you express your dissent with me upon some matters, and particularly relative to *potatoes*. The passage to which I allude, is in these words: "As to a former diatribe of his on potatoes, we regarded it as a pleasant example of argument for argument's sake; as an agreeable jumble of truth and of mental rambling."

Now, Sir, I do assure you that I never was more serious in my life, than when I wrote the essay, or rather, casually made the observations against the cultivation and use of this *worse than useless root*. If it was argument for argument's sake, no one that I can recollect ever did me the honour to *show* that the argument was fallacious. I think it a subject of great importance; I regard the praises of this root, and the preference given to it before corn and even some other roots, to have arisen from a sort of monkey-like imitation. It has become, of late years, the *fashion* to extol the virtues of potatoes, as it has been to admire the writings of Milton and Shakespear. God, *almighty* and all *fore-seeing*, first permitting his chief angel to be disposed to rebel against him; his permitting him to enlist whole squadrons of angels under his banners; his permitting this host to come and dispute with him the throne of heaven: his permitting the devils to bring cannon into this battle in the clouds; his permitting one devil or angel, I forget which, to be split down the middle, from crown to crotch, as we split a pig; his permitting the two halves, intestines and all, to go

slap, up together again, and become a perfect body; his,
then, causing all the devil host to be tumbled head-long
down into a place called Hell, of the local situation of
which no man can have an idea; his causing gates (iron
gates too) to be erected to keep the devil in: his permitting
him to get out, nevertheless, and to come and destroy the
peace and happiness of his new creation; his causing his
son to take *a pair of compasses* out of a *drawer*, to trace the
form of the earth; all this, and indeed, the whole of
Milton's poem, is such barbarous trash, so outrageously
offensive to reason and to common sense, that one is
naturally led to wonder how it can have been tolerated
by a people, amongst whom astronomy, navigation and
chemistry are understood. But, it is the *fashion* to turn up
the eyes, when Paradise Lost is mentioned: and if you fail
herein you want *taste*; you want *judgment* even, if you do
not admire this absurd and ridiculous stuff, when, if one
of your relations were to write a letter in the same strain
you would send him to a mad-house and take his estate. It
is the sacrificing of *reason* to *fashion*. And as to the other
"Divine Bard", the case is still more provoking. After his
ghosts, witches, sorcerers, fairies and monsters; after his
bombast and puns and smut, which appear to have been not
much relished by his comparatively rude contemporaries,
had had their full swing; after hundreds of thousands of
pounds had been expended upon embellishing his works;
after numerous commentators and engravers and painters
and booksellers had got fat upon the trade; after *jubilees*
had been held in honour of his Memory; at a time when
there were men, otherwise of apparently good sense, who
were what was aptly enough termed *Shakespear-mad*. At
this very moment an occurrence took place, which must
have put an end, for ever, to this national folly, had it not
been kept up by infatuation and obstinacy without parallel.
Young Ireland, I think his name was William, no matter
from what *motive*, though I never could see any harm in

his motive, and have always thought him a man most unjustly and brutally used. No matter, however, what were the inducing circumstances, or the motives, he did write, and bring forth, as being Shakespear's some *plays*, a *prayer* and *a love-letter*. The learned men of England, Ireland and Scotland met to examine these performances. Some *doubted*, a few *denied*; but, the far greater part, amongst whom were Dr. Parr, Dr. Wharton and Mr. George Chalmers, declared in the most positive terms, that *no man but Shakespear* could have written those things. There was a *division*; but this division arose more from a suspicion of some trick than from any thing to be urged against the merits of the writings. The plays went so far as to be *acted*. Long lists of subscribers appeared to the work. And, in short, it was decided, in the most unequivocal manner, that this young man of sixteen years of age had written *so nearly like Shakespear* that a majority of the learned and critical classes of the nation most firmly believed the writings to be Shakespear's and, there cannot be a doubt, that if Mr. Ireland had been able to keep his secret, they would have passed for Shakespear's 'till the time shall come when the whole heap of trash will, by the natural good sense of the nation, be consigned to everlasting oblivion; and, indeed, as folly ever doats on a darling it is very likely that these last found productions of *our immortal bard* would have been regarded as his *best*. Yet, in spite of all this; in spite of what one would have thought was sufficient to make blind people see, the fashion has been kept up; and what excites something *more* than ridicule and contempt, Mr. Ireland, whose writings had been taken for Shakespear's, was, when he *made the discovery*, treated as an imposter and a *cheat* and hunted down with as much rancour as if he had written against the buying and selling of seats in Parliament. The *learned* men: the *sage critics*: the *Shakespear-mad folks*; were all so *ashamed* that they endeavoured to draw the public attention from themselves to the young man. It was of *his impositions* that they now talked and not of their *own folly*. When the witty clown, mentioned in Don Quixote, put the nuncio's audience to shame by pulling the *real pig* out from under his cloak, we do not find that that audience were, like our *learned* men so unjust as to pursue him with reproaches and with every act that a vindictive mind can suggest. They perceived how foolish they had been, they hung down their heads in silence and I dare say, would not easily be led to admire the mountebank again.

It is *fashion*, Sir, to which in these most striking instances sense and reason have yielded; and it is to *fashion* that the potatoe owes its general cultivation and use. If you ask me whether fashion can possibly make a *nation* prefer one sort of *diet* to another, I ask you what it is that can make a *nation* admire Shakespeare? What is it that can make them call him a Divine Bard, nine-tenths of whose works are made up of such trash as no decent man now-a-days would not be ashamed, and even afraid, to put his name to? What can make an audience in London sit and hear, and even applaud, under the name of Shakespear, what they would hoot off the stage in a moment if it came forth under any other name? When folly has once given the fashion she is a very persevering dame. An American writer, whose name is George Dorsey, I believe, and who has recently published a pamphlet called *The United States and England etc.* being a reply to an attack on the morals and government and learning of the Americans, in the *Quarterly Review* states as matter of *justification* that the people of America sigh *with delight* to see the plays of Shakespear, whom they claim as *their countryman*; an honour, if it be disputed of which I make any of them a voluntary surrender of my share. Now, Sir, what can induce the American to sit and hear with delight the dialogues of Falstaff and Poins, and Dame Quickeley, and Doll Tearsheet? What can restrain them from pelting Parson Hugh, Justice Shallow, Bardolph, and

the whole crew off the stage? What can make them endure
a ghost cap-a-pie, a prince who for *justice* sake pursues his
uncle and his mother, and who stabs an old gentleman in
sport and cries out 'dead for a ducat! dead!' What can
they find to 'delight' them in punning clowns, in ranting
heroes, in sorcerers, ghosts, witches, fairies, monsters,
sooth-sayers, dreamers; in incidents out of nature, in scenes
most unnecessarily bloody. How they must be delighted at
the story of Lear putting the question to his daughters of
which loved him most and then dividing his kingdom among
them *according to their professions of love;* how delighted to
see the fantastical disguise of Edgar, the *treading out* of
Gloucester's eyes and the trick by which it is pretended he
was made to believe that he had actually fallen from the top
of the cliff! How they must be delighted to see the stage
filled with green boughs, like a coppice, as in Macbeth, or
streaming like a slaughter-house, as in Titus Andronicus!
How the young girls in America must be tickled with
delight at the dialogues in Troilus and Cressida and more
especially at the pretty observations of the Nurse, I think
it is, in Romeo and Juliet! But it is the same all through
the work. I know of one other and *only one other* book so
obscene as this; and, if I were to judge from the high favour
in which these two books seem to stand, I should conclude
that wild and improbable fiction, bad principles of morality
and politicks, obscurity in meaning, bombastical language,
forced jokes, puns and smut were fitted to the minds of the
people. But I do not thus judge. It is *fashion*. These books
are in fashion. Everyone is ashamed not to be in the
fashion. It is the fashion to extol potatoes and to eat
potatoes. Everyone joins in extolling potatoes, and all the
world like potatoes, or pretend to like them, which is the
same thing in effect.

In those memorable years of wisdom 1800 and 1801 you
can remember, I dare say, the grave discussions in Parliament about potatoes. It was proposed by someone to make

a *law* to encourage the growth of them; and if the Bill did not pass, it was, I believe, owing to the ridicule which Mr Horne Tooke threw upon that whole system of petty legislation. Will it be believed, in another century, that the law-givers of a great nation actually passed a law to compel people to eat pollard in their bread and that, too, not for the purpose of *degrading or punishing* but for the purpose of doing the said good people good by *adding* to *the quantity of bread* in a time of scarcity? Will this be believed? In every bushel of wheat there is a certain proportion of *flour*, suited to the appetite and the stomach of man; and a certain proportion of *pollard* and *bran*, suited to the appetite and stomach of pigs, cows and sheep. But the parliament of the years of wisdom wished to cram the *whole* down the throat of man, together with the flour of other grain. And what was to become of the pigs, cows and sheep? Whence were the pork, butter, and mutton to come? And were not these articles of human food as well as bread? The truth is, that pollard, bran and the coarser kinds of grain, when given to cattle, make those cattle fat; but when eaten by man make him lean and weak. And yet this bill actually became a law!

That period of wisdom was also the period of the potatoe-mania. *Bulk* was the only thing sought after; and, it is a real fact, that Pitt did suggest the making of *beer* out of *straw*. Bulk was all that was looked after. If the scarcity had continued a year longer, I should not have been at all surprised if it had been proposed to feed the people at rack and manger. But the *Potatoe*! Oh! What a blessing to man! Lord Grenville, at a birthday dinner given to the foreign ambassadors, used not a morsel of bread, but instead of it little *potatoe cakes*, though he had, I dare say, a plenty of lamb, poultry, pig, etc., all of which had been fatted upon corn or meal in whole or in part. Yes, Sir, potatoes will do very well along with plenty of animal food, which has been *fatted on something better* than potatoes.

But when you and I talk of the use of them, we must consider them in a very different light.

The notion is, that potatoes are *cheaper* than *wheat flour*. The word *cheap* is not quite expressive enough but it will do for our present purpose. I shall consider the *cost* of potatoes in a family, compared with that of flour. It will be best to take the simple case of the labouring man.

The price of a bushel of fine flour, at Botley, is, at this time, 10s. The weight is 56-lbs. The price of a bushel of potatoes is 2s. 6d. They are just now dug up and are the cheapest. A bushel of potatoes which are measured by a large bushel weighs about 60-lbs. dirt and all for they are sold unwashed. Allow 4-lbs. for dirt and the weights are equal. Well then, here is toiling Dick with his four bushels of potatoes and John with his bushel of flour. But, to be fair, I must allow that the relative price is not always so much in favour of flour. Yet I think you will agree with me that upon an average five bushels of potatoes do cost as much as one bushel of flour. You know very well that potatoes in London sell for 1d. and sometimes for 2d. a pound; that is to say sometimes for £1.7.6d. and sometimes for £2.15s. the five bushels. This is notorious. Every reader knows it. And did you ever hear of a bushel of flour selling for £2.15s. Monstrous to think of! And yet the tradesman's wife, looking *narrowly* into every halfpenny, trudges away to the potatoe shop to get five or six pounds of this wretched root for the purpose of *saving flour*! She goes and gives 10d. for ten pounds of potatoes, when she might buy five pounds of flour with the same money! Before her potatoes come to table they are even in *bulk* less than 5-lbs. or even 3-lbs. of flour made into a pudding. Try the experiment yourself, sir, and you will soon be able to appreciate the *economy* of this dame.

But to return to Dick and John; the former has got his five bushels of potatoes and the latter his bushel of flour. I shall, by and by, have to observe upon the *stock* that Dick

must lay in, and upon the stowage that he must have; but
at present we will trace these two commodities in their way
to the mouth and in their effects upon those who eat them.
Dick has got five bushels at once because he could have
them a little cheaper. John may have his *Peck* or *Gallon* of
flour; for this has a fixed and indiscriminating price. It
requires no trick in dealing, no judgment as in the case of
the roots, which may be *wet* or *hollow* or *hot*; flour may be
sent for by any child able to carry the quantity wanted.
However, reckoning Dick's trouble and time nothing in
getting home his five bushels of potatoes, and supposing
him to have got the *right* sort, a 'fine sort' which he can
hardly fail of indeed, since the whole nation is now full of
'fine sort', let us now see how he goes to work to consume
them. He has a piece of bacon upon the rack but he must
have some potatoes too. On goes the pot, but there it may
as well hang, for we shall find it a continual requisition.
For this time the meat and roots boil together. But what is
Dick to have for supper? Bread? No. He shall not have
bread unless he will have bread for dinner, Put on the Pot
again for supper. Up an hour before day light and on with
the pot. Fill your luncheon bag Dick: nothing is so relish-
ing and so strengthening out in the harvest-field or ploughing on a bleak hill in winter, as a cold potatoe. But be sure
Dick to wrap your bag well up in your cloathes during the
winter, or when you come to lunch you may to your great
surprise find your food transformed into pebbles. Home
goes Merry Dick and on goes the pot again. Thus 1095
times in the year Dick's pot must boil. This is at least a
thousand times oftener than with a bread and meat diet.
Once a week baking and once a week boiling is as much as
a farm house used to require. There must be some fuel
consumed in winter for warmth. But here are, at the least,
500 fires to be made for the sake of these potatoes, and, at a
penny a fire, the amount is more than would purchase four
bushels of flour, which would make 288 lbs. of bread, which

at 7 lbs. of bread a day would keep John's family in bread for 41 days out of the 365. This I state as a fact challenging contradiction, that exclusive of the extra *labour*, occasioned by the cookery of potatoes, the fuel required, in a year, for a bread diet, would cost in any part of the kingdom more than would keep a family even in a baker's bread, for 41 days in the year, at the rate of 71 lbs. of bread a day.

John, on the contrary, lies and sleeps on Sunday morning till about 7 o'clock. He then gets a bit of bread and meat or cheese, if he has either. The mill gives him his bushel of flour in a few minutes. His wife has baked during the week. He has a pudding on Sunday and another batch of bread before the next Sunday. The moment he is up he is off to his stable or the field or the coppice. His breakfast and luncheon are in his bag. In spite of frost he finds them safe and sound. They give him heart, and enable him to go through the day. His 56 lbs. of flour, with the aid of 2d. in yeast, bring him 72 lbs. of bread while, after the dirt and peelings and waste are deducted, it is very doubtful whether Dick's 300-lbs. of potatoes bring 200-lbs. of even this watery diet to his lips. It is notorious that in a pound of clean potatoes there are 11 ounces of water, half an ounce of earthy matter, an ounce of *fibrous* and *strawey* stuff, and I know what besides. The water can do Dick no good, but he must swallow these 11 ounces of water in every pound of potatoes. How far *earth* and *straw* may tend to fatten or strengthen cunning Dick I do not know; but at any rate, it is certain that while he is eating as much of potatoe as is equal in nutriment to 1-lb. of bread, he must swallow about 14 oz. of water, earth, straw etc., for down they must go altogether like the Parliament's bread in the years of wisdom, 1800 and 1801. But suppose every pound of potatoes to bring into Dick's stomach a 6th part of nutritious matter, including in the gross pound all the dirt, eyes, peeling and other inevitable waste. Divide this gross 300 lbs. by 6 and you will find him with 50-lbs of nutritious matter for the same sum that John has laid out in 72-lbs. of nutritious matter, besides the price of 288 lbs. of bread in a year, which Dick lays out in extra fuel for the eternal boilings of his pot. Is it any wonder that his cheeks are like two bits of loose leather, while he is pot-bellied and weak as a cat? In order to get half a pound of nutritious matter into him he must swallow about 50 ounces of water, earth and straw. Without ruminating facilities how is he to bear this cramming?

But Dick's disadvantages do not stop here. He must lay in his store at the beginning of winter, or he must buy through the nose. And where is he to find stowage? He has no caves. He may *pie* them in the garden, if he has one, but he must not open the pie in frosty weather. It is a fact not to be disputed that a full tenth of the potatoe crop is destroyed, upon an average of years by the frost. His wife, or stout daughter, cannot go out to work to help to earn the means of buying potatoes. She must stay at home to *boil the pot*, the everlasting pot! There is no such thing as a *cold dinner*. No such thing as women sitting down on a haycock, or a shock of wheat, to their dinner, ready to hump up at the approach of the shower. Home they must tramp, if it be three miles, to the fire that ceaseth not, and the pot as black as Satan. No wonder that in the brightest and busiest seasons of the year, you see from every cottage door, staring out at you as you pass, a smoky capped, greasy heeled woman. The pot, which keeps her at home, also gives her the colour of the chimney, while long inactivity swells her heels.

Now, Sir, I am quite serious in these my reasons against the use of this root as food for man. As food for other animals in proportion to its cost, I know it to be the *worst of all roots* that I know anything of. But that is another question. I have here been speaking of it as food for man; and, if it be more expensive than flour to the labourer in *the country*, who at any rate can stow it in pies, what must it be to tradesmen's and artizans' families in *towns*, who can lay in no store and who must buy by the ten pound or quarter of a hundred at a time. When broad-faced Mrs. Wilkins tells Mrs. Tomkins that so that she has a *potatoe* for her dinner, she *does not care a farthing for bread*, I only laugh knowing that she will twist down a half pound of beef with her 'potatoe' and has twisted down half a pound of buttered toast in the morning, and means to do the same at tea-time without prejudice to her supper and grog. But when Mrs. Tomkins gravely answers, 'yes, Ma'am, there is nothing like a potatoe, it is such a *saving* in a family,' I should not be very much out of humour to see the tete-a-tete broken up by the application of a broom-stick.

However, sir, I am talking to you now, and as I am not aware that there can be any impropriety in it, I now call upon you to show that I am really wrong in my notions upon this subject; and this I think you are in some sort bound to do, seeing that you have, in a public manner, condemned them.

I am, Sir,

Your most obedient

And most humble Servant,

Wm. COBBETT.

Now, observe I never received any *answer* to this. Much *abuse*. New torrents of *abuse*; and, in language still more venomous than the former; for *now* the Milton and Shakespear men, the critical *Parsons*, took up the pen; and when you have an angry *Priest* for adversary, it is not the common viper but the rattle-snake that you have to guard against. However, as no one put his *name* to what he wrote, my remarks went on producing their effect; and a very considerable effect they had.

I beg to be understood as saying nothing against the *culti-vation* of potatoes in any place, or near any place where there are people willing to consume them at *half a dollar a bushel*, when wheat is *two dollars a bushel*. If any one will buy *dirt* to eat, and

if one can get dirt to him with more profit than one can get
wheat to him, let us supply him with dirt by all means. It is
his *taste* to eat dirt; and, if his taste have nothing immoral in it,
let him in the name of all that is ridiculous, follow his taste.
I know a *Prime Minister* who picks his nose and regales himself
with the contents. I solemnly declare this to be true. I have
witnessed the worse than beastly act scores of times; and yet
I do not know that he is much more of a beast than the greater
part of his associates. Yet, if this were *all*; if he were chargeable
with nothing but this; if he would confine his *swallow* to this,
I do not know that the nation would have any right to interfere
between his nostrils and his gullet.

Nor do I say that it is *filthy* to eat potatoes. I do not ridicule
the using of them as *sauce*. What I laugh at is the idea of the
use of them being a *saving*: of their *going further* than bread; of
the cultivation of them in lieu of wheat *adding to the human
sustenance of a country*. This is what I laugh at; and laugh I
must as long as I have the above estimate before me.

As food for cattle, sheep or hogs this is the *worst* of all the
green and root crops; but of this I have said enough before;
and therefore I now dismiss the Potatoe with the hope that I
shall never again have to write the word or to see the thing.

Spinage. Every one knows the use of this excellent plant.
Pigs, who are excellent judges of the relative qualities of
vegetables, will leave cabbage for lettuces and lettuces for
spinage. Gardeners make two sorts of spinage, though I
really believe there is but one. One sort they call *round spinage*,
and the other *prickly spinage*, the former they call summer
spinage, and the latter winter: but I have sowed them indis-
criminately and have never perceived any difference in their
fitness to the two seasons of the year. The spinage is an annual
plant, produces its seed and ripens it well even if sowed so
late as the month of May. It may be as well to sow the round
spinage for summer, and the prickly spinage for winter, but
the time of sowing and the manner of cultivating are the only
things of importance; and great attention should be paid to

these, this being a most valuable plant all the year round, but particularly in the winter and the spring. It has something delightfully refreshing in its taste, and is to be had at a time when nothing but mere greens or broccoli is to be had. It far surpasses them both in my opinion, the use of it never being attended with any of those inconveniences as to bodily health which is the case with both the others. In the summer there are plenty of other things; but for the winter crop, due provision should always be made. The time for sowing for the winter crop, if the ground be good, is the last week in August, and if the ground be poor, a fortnight earlier. Sow in shallow drills, eights inches apart, and thin the plants to six inches apart in the row: keep them clear of weeds, hoe about them before winter sets in, and draw the earth close up to the stems of the plants, taking care that the dirt does not fall into the hearts. The ground should be rather of the drier description; for if wet and the winter be severe, the plants will be killed. They will have fine leaves in the month of November, or before: for use, the outside leaves should be taken off first, or rather these only should be taken off, leaving all the rest, and they should be pinched off with the finger and the thumb close to the stem of the plant. The plant will keep growing, more or less, all the winter, except in very hard weather, and will keep on yielding a supply from the beginning of November to the latter end of May, when the seed stalks will begin to rise, and when the summer spinage, sowed in the latter end of February and cultivated in the same way as the former, will be ready to supply their place. About the first of May, another sowing of summer spinage should take place; but this will be generally supplanted by peas, beans and other summer crops. If however the reader wish, like me, to have it all the summer, he must sow again in the month of June, and again in the month of July. These two latter sowings being made in the coolest and least sunny part of the garden. As to saving the seed of the spinage, a few plants of each sort will be sufficient. The plants must be pulled up before the seed be dead ripe, or the birds will have every grain. It is a coarse-looking seed with a thick husk upon it; but the small birds are very fond of it and will begin to hammer it out of the husks while these are still green. The seed-plants, when pulled up, should be laid in the sun to become perfectly dry, and the seed should be then rubbed off and put by in a dry place.

Tarragon is a very hot, peppery herb, used in soups and salads. It is perennial and may be propagated from seed sowed at any time in the spring or from offsets put out in either spring or fall. Its young and tender tops only are used. It is eaten with beef-steaks in company with minced shalots. A man may doubtless live very well without it; but an orthodox clergyman once told me that he and six others once ate some beef-steaks with shalots and tarragon and that they 'voted unanimously that beef-steaks never were so eaten.' If you will have it in winter, you must dry it in the manner directed for sage and other herbs.

Nota Bene – *Borage.* I omitted the insertion of this plant in due alphabetical order, and as the printer treads closely upon my heels, I am obliged to mention it here. This is a very pretty flowering plant. One sort of it has *blue* flowers, one *red*, and another *white*. The only use that I ever saw borage put to, was putting it into wine and water along with nutmeg, and some other things perhaps, the mixture altogether being called *cool-tankard*, or by the shorter name *cup*. If once you have it growing upon any spot you need not take the trouble to sow it. It bears an abundance of seed, some of which is ripe while the plant is still in bloom. If you wish to have it young at all times, you may sow in the spring, in the summer, in autumn or at any time. The plants should not stand too thick upon the ground, and the ground should be kept clean. Any awkward corner under one of the hedges will do very well for borage, which, however, is by no means unornamental in a flower-garden, both flower and leaf being very pretty.

GARDEN SEEDS

I, some time ago, notified my intention of selling garden seeds
this winter; and I am now prepared to do it. Those who have
read my writings on Agriculture and Gardening, and particularly my *English Gardener*, will have perceived that I set
forth, with much pains, the vast importance of being extremely
careful with regard to the seeds which one sows: and, as to
which matter, there are two things to be attended to; first, the
genuineness of the seeds; and next, as to their *soundness*. The
former is the more important point of the two; for it is a great
deal better to have no plants at all, than to have things come
up, and, at the end of a month or two, to find that you have got
a parcel of stuff, not at all resembling that which you thought
you were about to have. Those who have read my *Gardening
Book, chap.* 4, will want very little more to convince them of the
importance of this matter. I have always taken great delight in
having perfect plants of every description; but, to get into the
way of raising good and true garden seeds, requires that you
be settled upon some sufficient space of ground for *several suc-
cessive years*; and it has been my lot to live under a Government,
which, if you take the liberty to differ from it in opinion, has
taken care to prevent you, by hook or by crook, from being
settled in any place, except one of its prisons, for any considerable length of time. Since, however, it, in a lucky hour,
had the wisdom to pass Peel's Bill, it has been rather less
rummaging; though it certainly *meant well* towards me in the
year 1831. I have, however, been suffered to remain long enough
at *Kensington* to bring the seeds of a good many plants to what
I deem perfection, and others nearly to that state; and I have
taken a little farm in Surrey, partly for the purpose of raising
garden seeds upon a greater scale than I was able to do it at
Kensington; this year I have raised a considerable quantity of
seeds, which I now offer for sale in the following manner, and
on the following terms.

It does not suit me to keep a *seedshop*, and to retail seeds by

the small quantity; but to make up packages, each sufficient
for a garden for the year, and to sell that package for a fixed
sum of money. When I was driven to *Long Island* by Sidmouth's
dungeon bill, and when the Hampshire parsons and Sidmouth
and Castlereagh, chuckled at the thought of my being gone to
mope away my life in melancholy, in the United States; and
when the famous traveller, Mr. Fearon brought home word,
that I was whiling away my life in a dilapidated country house,
the paths to which were over-run with thistles and brambles;
when Mr. Fearon, that accurate observer, exclaimed, in the
language of his brother Solomon, "Lo! it was all grown over
with thorns, and nettles covered the face thereof, and the post
and rail fence thereof was broken down;" when Mr. Fearon,
in the fulness of his compassion, was thus exclaiming, I, though
he found me in a pair of Yankee trousers not worth a groat,
was preparing to sell seeds in a house at New York, for which
I gave fourteen hundred dollars a year. In short, I imported a
a great quantity of seeds from London, which I sold principally
in the following manner:

I had *little boxes* made, into each of which I put a sufficiency
of each sort of seeds for a gentleman's garden for the year. The
large seeds were in paper bags, and the smaller seeds in papers.
In the box along with the seeds, I put a printed paper containing a list of the names of the several seeds, and against
each name the *number*, from numbers *one* to the end: then, there
were corresponding numbers marked upon the bags and the
parcels. So that, to know the sort of seed, the purchaser had
nothing to do but to look at the numbers on the parcels and
then to look at the list. Many of these boxes of seeds went as
far as *Lower Canada* to the north, and into the *Floridas,* to
New Orleans and even to the West India islands, to the south;
and the net proceeds were amongst the means of enabling me
to prance about the country; amongst the means of enabling
me to lead a pleasant life; of enabling me to stretch my long
arm across the Atlantic, and to keep up the thumping upon
Corruption, which I did to some tune.

I intend to dispose of my seeds in the same manner now, except that I shall use *coarse linen bags* instead of boxes. The several parcels of seeds will be put up either in paper bags or paper parcels; and a printed list with the *names* and *numbers* will be prepared; and then, the parcels and the list will be put into the linen bag, and sewed up, and will be ready to be sent away to any person who may want it.

A bag for a considerable garden; a garden of the better part of an acre, perhaps, will be sold for *twenty-five shillings*; and for a smaller garden, for *twelve shillings and sixpence*. These seeds, if bought at the shop of a seedsman, would come to more than three times the money; and so they ought: for the seedsman has his expensive shop to keep; has his books to keep; has his credit to give, and has his seeds to purchase with his ready money. While, therefore, I have a right to proceed in my manner, he does nothing wrong. By the lists, which I publish below, the reader will perceive that, to the garden seeds I have added the seeds of several annual flowers. They are not of very *rare* kinds; but they are all very pretty; and, even these flower seeds alone, if purchased at a seedsman's, would come, and ought to come, to pretty nearly one half of the money which I charge for the whole. Of some of the sorts of seeds the purchaser will think the quantity *small*; and, of these the cauliflower is one; but, it must be a thundering garden that requires more than three hundred cauliflower plants; and, if carefully sowed, agreeably to the directions in my *Gardening Book*, the seed which I put up is more than sufficient for any gentleman's garden; and I will pledge myself for the soundness of every individual seed. In the *small bag*, the quantity is in proportion to the price. Authors always want people to read their books; or, to purchase them at least. The reader will not, therefore, be surprised, that I *most earnestly exhort* all those who buy my seeds, *to buy my book, too*, and even then they will not have half so much to pay as if they had to purchase the seeds of a seedsman.

I have only one fear upon this occasion, and that is, that gentlemen's gardeners, who are in the habit of dealing with seedsmen, and who are apt to adhere too literally to that text of Scripture, which says that 'he who *soweth abundantly* shall reap abundantly;' but, begging their pardon, this does not mean covering the ground with the seeds, which, though it may produce abundant reaping to the seedsman, is far from having that tendency with regard to the crop. *Thick sowing* is, indeed, injurious in three ways: first, it is a waste of seed and of money, of which it is actually a flinging away of both: second, it makes work in the thinning out of the plants: third, the plants will never be so fine if they come up thick. Therefore, in my *Gardening Book, chapter* 4, beginning at paragraph 85, I take very great pains to give instructions for thin sowing; and, if every one who cultivates a garden could see the regularity, the cleanness, and the beauty, of my seed beds, never should we again see a parcel of seeds flung promiscuously over the ground. It is probable, that three hundred cauliflower seeds will lie in a thimble; and if you want three hundred plants, it is better to sow these three hundred seeds in a proper manner, than to fling twenty thousand seeds over the same space of ground. You must cut the superfluous seeds up with a hoe, or pull them out with your hand; and, small as they are, and insignificant as you may think their roots to be, they rob and starve one another, even before they get into rough leaf. I know very well, that it requires a great deal *more time* to sow a bed of a hundred feet long, for instance, and with cabbages, for instance; a great deal more time to sow it in drills, and to put the seed in thinly, than to fling the seed thickly over the ground and just rake it in; but, look at the *subsequent operations*; and you will find that, in the end, this '*sowing abundantly*' costs ten times the time and the labour which are required by the method of sowing pointed out in my book. Therefore, let no man imagine, that to have a plentiful crop a great quantity of seed is necessary. When, indeed, you have reason to fear that the seed is not sound, and when you cannot obtain that which you know to be sound, it may be prudent to throw in great parcels of it in order to have

the best chance to get *some* plants; but, hap-hazard work like
this ought to be avoided, if possible; and, at any rate, I pledge
myself, for the soundness of all my seed; I pledge myself that,
if properly sowed, every seed that I sell shall grow. Thus far
as to my seeds in general. I have now to speak of one sort of
seed, which, as that horrible old Whig, Sir Robert Walpole,
said of his bribes, '*is sold only at my shop.*' This is the seed of the
Cisalpine strawberry: this strawberry, unlike all others that I
ever heard of, *produces its like* from the seed; is raised with the
greatest facility, bears most abundantly, and *keeps bearing
until the hard frosts come.* The seeds are so small that a little
pinch of them between the finger and the thumb is sufficient
for a very large garden; and the method of rearing the plants
is this: about the first week of February, or it may be a little
later, fill with fine earth, to within about an inch of the top,
a flower-pot from twelve to fifteen inches over; take the little
pinch of seed and scatter it very thinly over the top of the earth;
then put some very fine earth over the seed a quarter of an
inch thick, or rather less. Set the pot in a green-house,
or in the window of any room where the sun comes,
and give water very carefully, and very gently, as
occasion may require. When the warm weather comes, the
pot should be set out of doors in a warm place when there is no
heavy rain, and should be taken in at night if there be any fear
of frost. Towards the end of *April,* the pot may be set out of
doors altogether; and, small as the plants will still be, they will
be fit to be planted out in the natural ground by the middle, or
towards the latter end, of *May.* Then dig a piece of ground deep,
and make it extremely fine upon the top, and put out the little
plants in rows *two feet apart*, and *two feet* apart in the row; for,
though not bigger than a thread, each plant will multiply
itself into a considerable *tuft* before the middle of July; and
then they will begin to bear, and they will keep on bearing as
long as the hard frosts keep away. The very runners which
proceed from these plants, will take root, blow and have ripe
fruit, during the first autumn. When the bearing is over, cut

off all the runners, clear the ground close up to the *tufts*, and let the tufts remain to bear another year, when their produce is prodigious. But, then you must grub them up; for they so multiply their offsets, and so fill the ground with their roots, that they almost cease to bear if they remain longer. So that you must have a new plantation from seed every year; and the seed you may save yourself, by squeezing the pulp of dead-ripe strawberries in water, which sends the seed to the bottom of the water; you skim off the pulp, and drain away the water, then put the seed out in the sun to dry, and then put it up and preserve it for sowing in the winter. There is a *red* sort and a *white* sort, which you may keep separate or sow them and plant them promiscuously. And, now, to do justice to Sir Charles Wolsley, who is my teacher as to this piece of knowledge, and at whose house, at *Wolsley Park*, I saw, in September last, the finest dishes of strawberries that I ever had seen in the whole course of my life. They were served up in a mixed state, some red and some white; and the taste and fragrance were equal to the beauty. Sir Charles was so good as to make his gardener save me a considerable quantity of the seed, which, by the bursting of the paper, became mixed; and, therefore, the parcels of this strawberry seed, which I shall put into my packages, will, the purchaser will bear in mind, be some of the *white* strawberry and some of the *red*. After this long story about garden seeds, which, however, is not so execrably stupid as the impudent babble of the Whigs about having '*settled upon a Speaker* for the next House of Commons', I proceed to give a list of the names of my seeds, and of the numbers which are to be put upon the parcels; once more observing, that a large package of seeds will be sold for *twenty-five shillings*, and a small one for *twelve shillings and sixpence*. A direction may be sewed on the package in a minute, and it can be sent to any part of the country by the coach, or in any other manner, as the weight, even of the larger package, is only about 16 pounds.

KITCHEN GARDEN SEEDS

1. Asparagus
2. Bean – Broad, or Windsor
3.   Long-pod
4.   Early Masagan
5.   Kidney (or French)
             Scarlet Runners
6.   White Runners
7.   Black Dwarf
8.   Dun Dwarf
9.   Robin-Egg
10.   Speckled
11. Beet – Red
12. Brocoli – White
13.   Purple
14. Cabbage – Early Battersea
15.   Early York
16.   Savoy
17. Cale – Curled – Scotch
18. Carrot
19. Cauliflower
20. Celery
21. Chervil
22. Cress

23. Cucumber, early frame
24. Corn (Cobbett's)
25. Endive
26. Leek
27. Lettuce – White Coss
28   Russia Coss
29.   Brown Dutch
30.   Green Cabbage
31. Mustard – White
32. Nasturtium – Dwarf
33. Onion
34. Parsnip
35. Parsley – Curled
36. Pea – Early-frame
37.   Tall Marrowfats
38.   Dwarf Marrowfats
39. Radish – Early Scarlet
40.   White Turnip
41. Spinage
42. Squash (from America, great variety)
43. Strawberry – Cisalpine
44. Turnip – Early-Garden

FLOWER SEEDS

45. Canterbury Bells
46. Catch Fly
47. China-asters
48. Clarkia, (very beautiful)
49. Convolvulus – Dwarf
50. Indian Pink
51. Larkspur – Dwarf Rocket
52. Lupins – Dwarf Yellow
53. Marvel of Peru
54. Poppy – Carnation

55.   French
56. Stock – White Wall-flower
57.   Scarlet, ten-week
58. Mignonette
59. Sweet-William
60. Sweet Pea
61. Venus's Looking-glass
62. Virginia Stock
63. Wall-flower

FIELD SEEDS

Swedish Turnip Seed – Any quantity under 10lbs. *9d.* a pound; and any quantity above 10lbs. and under 50lbs. *8d.* a pound; any quantity above 50lbs. *9d.* a pound; above 100lbs. *7d.* A parcel of seed may be sent to any part of the kingdom; I will find proper bags, will send it to any coach or van or wagon, and have it booked at my expense; but *the money must be paid at my shop before the seed be sent away*; in consideration of which I have made due allowance in the price. If the quantity be small, any friend can call and get it for a friend in the country; if the quantity be large, it may be sent by me.

Mangel Wurzel Seed – Any quantity under 10lbs. *8d* a pound; any quantity above 10lbs. and under 50lbs. *7d* a pound; any quantity above 50lbs. *6d.* a pound; any quantity above 100lbs. *6d.* a pound. The selling at the same place as above; the payment in the same manner.

TREE SEED

Locust Seed – 6s. a pound.

TREES AND TREE PLANTING

The inducements to create property by tree-planting are so many and so powerful that to the greater part of those who possess the means, little, I hope, need be said to urge them to the employing of those means. Occasions enough will offer for showing how *quickly* the profits come. But, still there are some persons, who possess such means, who are well assured of the ultimate gain, but who are, nevertheless, discouraged by the thought that they shall *not live* to see the actual pecuniary product of their undertaking, and who, according to the idea of the dismal moralist, Dr. Johnson, begin *to think of dying* when they are exhorted to plant a tree. Let all such attend to the lesson given them in La Fontaine's beautiful fable of the *Old Man and the Three Young Men*, the wise, the generous, the noble sentiments of which ought to be implanted in every human breast.

Un octogénaire plantoit
Passe encor de bâtir; mais planter à cet âge
Disoient trois jouvenceaux, enfants du voisinage
Assurément il radotoit
Car, au nom des dieux, je vous prie,
Quel fruit de ce labeur pouvez-vous recueillir?
Autant qu'un patriarche il vous faudroit viellir
A quoi bon charger votre vie
Des soins d'un avenir qui n'est pas fait pour vous?
Ne songez désormais qu'à vos erreurs passées
Quittez le long espoir et les vastes pensées;
Tout cela ne convient qu'à nous.
Il ne convient pas à vous-mêmes
Repartit le viellard. Tout établissement
Vient tard et dure pen. La main des Parques blêmes
De vos jours et des miens se joue également
Nos termes sont pariels par leur courte durée.
Qui de nous des clartes de la voûte assurée
Doit jouir le dernier? Est-il aucun moment
Qui vous puisse assurer d'un second seulement?
Mes arrière-nereux me devront cet ombrage:
Hé bien! défendez-vous au sage
De se donner des soins pour le plaisir d'autrui?
Cela même est un fruit que je goûte aujourd'hui
J'en puis jouir demain, et quelques jours encore;
Je puis enfin compéter l'aurore
Plus d'une fois sur vos tombeaux.
Le viellard eut raison; l'un des trois jouvenceaux
Se noya dès le port, allant à l'Amérique;
L'autre, afin de monter aux grandes dignités
Dans les emplois de Mars servant la république
Par un coup imprévu vit ses jours emportés;
Le troisième tomba d'un arbre
Que lui-même il voulut enter;
Et pleures du vieillard, il grava sur leur marbre
Ce que je viens de raconter.

To *translate* this is like an attempt to make a thing resemble the Rainbow; and, therefore, I beg those who may happen not to understand French, to be pleased to receive, from my pen, the following statement of the more prosaic meaning of these words of this absolutely inimitable writer, who, in marks of simplicity the most pleasing that ever followed the movements of a pen, has, on numerous subjects, left, to ages unborn, philosophy the most profound and sentiments the most just and exalted.

A man of fourscore was planting trees. 'To build might pass, but to plant at such an age!' exclaimed three young men of the neighbourhood. 'Surely,' they said, 'you are doting; for in God's name, what reward can you receive for this, unless you were to live as long as one of the Patriarchs? What good can there be in loading your life with cares about a time which you are destined never to see? Pray devote the rest of your life to thoughts on your past errors, give up distant and grand expectations: these become only us young men' – 'They become not even you,' answered the old man. 'All we do comes late, and is quickly gone. The pale hand of fate sports equally with

your days and with mine. The shortness of our lives puts
us all on a level. Who can say which of us shall last
behold the light of heaven? Can any moment of your lives
secure you even a second moment? My great grandchildren
will owe shady groves to me: And do you blame me for
providing delight for others! Why, the thought of this is,
of itself, a reward which I already enjoy; may I enjoy it
more tomorrow, and for some days after that; nay, I
may more than once even see the sun rise on your graves.'
The Old Man was right: one of the three, ambitious to see
the New World, was drowned in the port; another, pursuing fame in the service of Mars, was suddenly stopped by
an unexpected shot; the third fell from a tree, on which he
himself was putting a graff: and the Old Man, lamenting
their sad end, engraved on their tomb the story here
related.

SOME TREES

Common English Ash: Fraxinus Excelsior. It is well known that
the Ash grows to a very great height, and that it will,
if left to grow, become a very large tree. It is also well
known, that it is a beautiful tree. Gilpin calls it the
Venus of the woods. It has, however, one great disadvantage;
that is, that it puts on leaves later in the spring, and loses them
earlier in the fall, than any other English tree. But perhaps
Gilpin was thinking of a *naked* Venus, and then, indeed, the
Ash claims the pre-eminence in our woods. Laying aside this
nonsense, however, of poets and painters, we have no tree of
such various and extensive use as the Ash. It gives us boards;
materials for making implements of husbandry; and contributes towards the making of tools of almost all sorts. We could
not well have a waggon, a cart, a coach or a wheelbarrow, a
plough, a harrow, a spade, an axe or a hammer, if we had no
Ash. It gives us poles for our hops; hurdle gates, wherewith to
pen in our sheep; and hoops for our washing-tubs; and assists
to supply the Irish and West Indians with hoops for their pork

barrels and sugar hogsheads. It therefore demands our particular attention; and from me, that attention it shall have.

As to the soil, no tree that I know of, except the Birch, is so little choice as the Ash. On dry ground, on wet ground, on sand, on clay, on chalk, and on almost a swamp, if it be not quite filled with water, the Ash will grow and thrive, if it have anything like fair treatment. It has another quality which must render it particularly suitable to exposed situations, namely, that it fears not the winds. I do not mean the power which the winds have in bringing it down; though it is pretty stout in that respect, and does not come trembling about like an Elm before every south-western puff. I mean that sort of power which the winds have in checking the growth of trees, and especially the winds near the sea coasts. On the Hampshire coast, the wind that comes from the Atlantic is, of course, a south-west wind. You will see the oaks when exposed to this wind, shaved up on the south-west side of them, as completely as if shaven with a pair of shears. The head of the tree resembles the top of a board quickset hedge, which is kept sheared up in a sloping form on one side only. The head of such oak is still more, perhaps, like the roof of any semi-circular building clapped up against a house; and at the same time, the everlasting flinching of the tree, and the continuance of the weight on one side, while it is kept shaven off on the other, makes the trunk of the tree lean away from the south-west. Close by the side of an Oak like this you will see an Ash, of equal height and size, standing as upright as if in the warmest of valleys; and I have looked, with the most scrutinizing eye, without ever having been able to discover that any of the shoots pointing to the south-west, had received the smallest injury. I have seen thousands, and thousands of instances of this hardiness in the Ash, and a very valuable quality it is, though this astonishing hardiness is, it must be confessed, but little consonant with Mr Gilpin's ideas. He called the Oak, the 'Hercules of the Woods;' and, as I have shown, this Hercules flees at the bare approach of that which the Venus sets at defiance.

The Locust*. The outward appearance of this tree, its beautiful
leaves and flowers, are pretty well known in most parts of
England; but it remained for me to make known the properties
of the wood. These properties, too, are in parts mentioned by
Miller: and its surprising powers, when constituting parts of
ships, are mentioned at full length in Hunter's edition of
Evelyn's *Sylva*. Notwithstanding this, we never hear of a man
in England that ever planted this tree, until I took the matter
in hand, except as a thing of mere ornament, in which respect
it certainly surpasses any other in the world, but as such I
should not have deemed it worthy of notice. I have, at different times, written and published upon this subject, through
the channel of the *Register,* in which I began by producing
certificates relative to the durability of the wood. I shall, by
and by, publish those certificates, which I collected, or, at
least, I paved the way for collecting, while I was in Long
Island, from the month of May, 1817, to the month of October,
1819; a space of time that I was in voluntary banishment, for
the purpose of avoiding those dungeons into which such numbers
of the public-spirited and virtuous reformers were put, de-
prived of the use of pen, ink and paper, and from which they
were finally released (those of them who survived their sufferings) without any charge having been preferred against
them from the first to the last. Nevertheless, I did not forget
my country, and the duty I still owed to her. I was convinced
that nothing in the timber way could be so great a benefit as
the general cultivation of this tree. Thus thinking I brought a
parcel of this seed home with me in 1819 but I had no means
of sowing it until the year 1823. I then began sowing it, but
upon a very small scale. I sold the plants; and, since that time,
I have sold altogether more than a million of them.

The wood is very hard, and close and heavy; it is yellow
almost as box, as hard as box, but the grain not so fine. The
durability of this wood is such, that *no man in America will pretend*

* Acacia

to say that he ever saw a bit of it in a decayed state. This seems hyperbolical; but every American of experience in country affairs will, if appealed to, confirm what I say. It is absolutely indestructible by the powers of earth, air and water. Its strength far surpasses that of the very best of our Spine Oak. It is to this timber that the American ships owe a great part of their superiority to ours. The stantions round the deck are made of Locust; and, while much smaller than the stantions of Oak, will resist a sea three times as heavy as the Oak will. The tiller of the ship is made of Locust, because it demands great strength and is required not to be bulky. For the same reason, the martingales of ships are made of Locust. The Locust is rather a rare timber in America; but sometimes the *Fettocks* or *ribs* of ships are made in Locust; and if a ship had all its ribs and beams and knees of Locust, it would be *worth two common ships.* Further, as to ship-building, that important article, the Trunnels, when they consist of Locust, make the ship last, probably twice as long as if the trunnels consisted of Oak. Our Admiralty know this very well, or at least they ought to know it. These trunnels are the *pins,* of which so many are used to hold the side-planks on to the timbers of the ship. Trunnels is said to be a corruption from tree-nails; but I do not believe it. However, we know what these things are: we know that they are an article of the very first importance in ship-building; we know that the *hardest of our Spine Oak* is picked out for the purpose; and, with all that, we know that the trunnel is the thing that rots first: for the water, or, at least, the damp, will get in round the trunnel, and between it and the plank; and if water or damp hang about Oak, the Oak *will rot.* All the American *public* ships are built with Locust trunnels, and so are all the merchant-ships of the first character.

Some of our own public ships have, I fancy, Locust trunnels brought from America; and I have been informed, that when Cropper, Benson, and Co., of Liverpool, built their East Indiamen, they imported the Locust trunnels and some other of the timbers from New York. We have a monstrous deal to do, in many respects, to make our navy (gun for gun) a match for that of the United States; but if we had accomplished every other point, there would remain want of timber, unless we supplied ourselves with Locust, at the least. The Hickory we should want for handspikes, for mast-hoops and other hoops to go round the yards and stays. Various other things would be wanted to make our ships as light and as roomy as those of the Americans, and with the same degree of strength; but, without the Locust, it is impossible to match them.

But, important as these matters are, these are, by no means, to be compared to the various uses about *buildings* and fences. I have said that this wood is indestructible, by the elements, except that of fire. How many thousands of houses are rendered useless in England, every year, by that thing which they call the dry rot, proceeding solely from those villanous soft woods which impatient people take such delight in planting and which carpenters of delicate constitution take such delight in sawing and planing. English Spine Oak is stronger than Deal; and if you keep it dry, it will not rot; but let it lie in the wet, or damp, and let the air get at it at the same time, and no villanous Deal board will turn to earth more quickly. Window sills of the best of Oak will rot, if something be not done to keep away the wet from getting under them: and in this very way the dry rot has got into many a house. Oak *door* sills are rotten in a very short time. The ends of beams and of joists, if they rest upon brick or stone, where the moisture is constantly about them, rot in a few years. The points of rafters, and the pins which hold rafters together, are always rotting. If these things were made of Locust your house would be safe for ages. Everywhere, when you want something to lie sopping in the wet, and at the same time to be exposed to the air, you should have Locust. Endless are the uses to which it might be put. A bottle-rack, for instance, that you want to stand out of doors and hidden in some corner, a grindstone-stand, a horse-block; but particularly, a cart-house, or anything that requires *pillars*, the *bottoms* of which are to go *into the ground*. Go into any farm-yard in England; I do not care what farm-yard it is and you shall find, in the cart-house, one of these things: first, the posts that support the buildings rotting off very fast, just where they meet the ground; second, those posts rotted off and cut *off*, and some stones put under them, to the manifest risk of the cart-house; third, the cart-house actually tumbling down in consequence of the rotting off of the posts. This is notorious; every farmer, every landlord in the kingdom, knows it. Now, I will insert a note from my memorandum-book, under date of October 16, 1819. 'At Judge Lawrence's, at Bayside, I saw a new cider-house, built against a hill, the upper story of it supported in front by some Locust posts. These posts, the Judge told me, had stood for *forty years*, or rather better, as the posts of a *cart-shed*." They were as sound as they had been the first year they were cut down. In our stables in England, you see stones put at the bottom of the stall-posts. What a plague it is! Little Locust trees, only about seven years old, would, for these purposes, make posts that would last for ever. Everyone knows how the sleeper (as I think they call it) rots; that is to say, the piece of wood that goes along at the bottom of each side of the stall. We know, also, how the manger-posts rot off at the ground. Use Locust timber, and it will sear out the stone walls of the building.

I should fatigue the reader were I to enumerate only a tenth part of the uses of this timber; but in short, if the timber be *imperishable*, what need of anything more in its praise? Will, however, English people *believe* in the imperishability? I would not believe in such a thing, if no *proof* were produced; and, therefore, I will now proceed to the proof of the truth of what I have stated. The test of imperishability is the situation of the *post* or *sill* being exposed to air and water; or rather, it being so situated as to *lie sopping in the wet*; I was led, by circumstances to be stated by and by, to entertain, while I was last in America, an anxious desire to introduce this valuable tree into England. After I had resolved to return in 1819, I set myself to work to get some seed together, which I found to be

no easy matter; for the Locust tree is by no means abundant in
any part of America where I have lived; but, how to go to
work to persuade English people that a little tree, chopped
down, and put into the ground as a gate post or pale post,
would stand there for a hundred years without rotting at all!
How to persuade English people to believe this; and to believe,
of course, that there was a timber about a hundred times as
good as their heart of Oak! You shall hear how I went to work
to endeavour to effect this.

In the latter end of August, in the year just spoken of, I was
at Plandome, the farm and residence of Mr. Judge Mitchell in
Long Island. He was building a new house on the spot where
had stood the house of his grandfather. There had been a little
sort of lawn before the door, enclosed by a pale fence. The
fence had all been pulled up, and there it lay, posts and rails
and pales. I asked the Judge how long the posts had been in
the ground. He said *eight and twenty years*. Each post had been
a little tree, just chopped down, sawed off to the proper length,
and squared, and each containing about *half a foot of timber*.
They were all as sound as they had been the first day that they
were cut down; and even the little sharp edges left by the *axe-chops*,
at the part where the square part met with the unsquared part;
even the little axe-chops were sound. The Americans use what
they call *stakes*, to hold on the top-rail of what they call a
worm-fence. These are generally made of little limbs of trees,
about eight feet long, and about the bigness of a hop-pole. I
saw many of these at Judge Mitchell's on that day, which he
assured me had been standing as stakes for upwards of *thirty
years*. I hinted to the men of Kent that I would teach them how
to make everlasting hop-poles; and this is a duty that I particularly owe to my native town of Farnham, so famous for
hops.

On the 25th of October of the same year, 1819, I was in
company with Doctor Peter Townsend, at Mr. Judge Lawrence's
at Bayside, in the township of Flushing, Long Island. I was talking to them about this Locust-tree project; and here I cannot

refrain from making an observation; which I have more than
once made in my *Year's Residence*;* namely, that, say, what they
will of the selfishness of Jonathan, I say that he is the most truly
liberal of all mankind. At home he never grudges his neighbour
his good fortune; he is always made happy by his neighbour's
success and prosperity; and, as to foreign nations, he is always
anxious that they should possess all the products, all the inventions, all the improvements that he himself enjoys. In con-
formity with his most amiable disposition, my excellent friends
at Bayside entered into my views about introducing the Locust
to England. The Judge showed me a post, which he said must
be nearly a hundred years old as a post. This post had been cut
down, when a little tree, and it had served in the capacity of
what they call a hog-gallows post. I examined it very minutely
and I found it perfectly sound, even to the very tips of it. It was
a post with a fork at the top of it. The points of the fork had
been chopped off in a careless manner; and there were these
points perfectly sound. But the main question was, *how was the
post where it met the ground*? It was just as good there as it was in any
other part. It had stood in a *gutter*, observe, for all these number
of years. The water thrown to wash out the hogs had run down
the gutter and had soaked down about the post. The numerous
sweepings and shovellings of the gutter, to take away the blood
and the mud, had *worn* away the post a little, as they would
have worn away iron; but still it was as sound as on the day
when it was *felled*.

Judge Mitchell was so kind as to give me a memorandum,
signed by himself, relative to this post; and Judge Lawrence
not being so old as his brother Henry, we sent for the latter and
he signed a memorandum, relative to the hog-gallows post. I
dare say that every reader, who delights in rural concerns, and
who duly considers the vasts importance of this matter, will
lament that he also could not see these posts. If he happen to

* *Journal of a Year's Residence in the United States of America*
(*1818–19*)

be in London, *he may see them now*; for they are to be seen by anybody at the Office of the *Register* in Fleet Street. The Manchester Magistrates brought out horse and foot to prevent me from passing through their town. The Bolton Magistrates put John Hayes in prison for ten weeks for announcing that I had arrived at Liverpool in good health. But my Locust posts came safely to London, and I came soon after them, with the following memorandums in my pocket.

Plandome, 23 August 1819

I have this day given to William Cobbett a Locust post six feet long, and squaring three inches by three and a half, which is perfectly sound in all its parts, and which has stood in the ground, as part of a fence, in front of my house, from the year 1791 until about five weeks ago, when the fence was taken up.

Singleton Mitchell.

Bayside, Flushing,
25 Oct. 1819.

My Brother, Effingham Lawrence, has this day taken up out of the ground, and given to William Cobbett, a hog-gallows post; that is, a post having a fork at the top, for the purpose of lodging a pole on, and on which pole hogs are, when killed, hanged up by the heels. This post is of Locust Wood; it was a single tree, and the whole of the lower part of that tree; it is, from extreme point to extreme point, eight feet eight inches long; from the tip of one fork to that of the other, from outside to outside, is seventeen and a half inches; there is a knot, the middle of which is fourteen and a half inches from the end of the butt; there is another knot eleven inches from the middle of the fork; the circumference of the post, at the mid-distance from the ends, is eighteen inches. I have known this post standing as a hog-gallows post during *forty-four years*. When I first knew it, it was a very old post. I remember hearing my father say that it was a *wonderfully old* post then. I should suppose it to have been a post upwards of *fourscore years.*

Henry Lawrence.

The Thorn (Black). The tree, or rather the bush, on the subject of which I am now about to hope for the reader's attention, is pretty well known to most English people, who will generally, perhaps, look upon it as something of little importance; but which is of real importance as to the two great purposes to which it is applied; namely, the making of excellent *hedges*, and the making of excellent *Port wine*: in which last of its functions I shall consider it first.

Everyone knows that this is a Thorn of the Plum kind; that it bears very small black plums, which are called Sloes, which have served love-song poets, in all ages, with a simile whereby to describe the eyes of their beauties, as the snow has constantly served them with the means of attempting to do something like justice to the colour of their skins and the purity of their minds, and as the rose has served to assist them in a description of the colour of their cheeks.

These beauty-describing sloes, have a little plum-like pulp, which covers a little roundish stone, pretty nearly as hard as iron, with a small kernel in the inside of it. This pulp, which I have eaten many times when I was a boy until my tongue clove to the roof of my mouth and my lips were pretty nearly glued together, is astringent beyond the powers of alum. The juice expressed from this pulp is of a greenish black, and mixed with water, in which a due proportion of logwood has been steeped, receiving, in addition, a sufficient proportion of cheap French brandy, makes the finest Port wine in the world, makes the whiskered bucks, while they are picking their teeth after dinner, smack their lips observing that the wine is beautifully rough and that they like 'a *dry wine* that has a good *body*'.

It is not, however, as a fruit-tree that I am here about to speak seriously to sensible people; it is of a *bush* excellent for the making of *hedges*, and not less excellent for the making of

walking sticks and swingles of flails. The Black Thorn blows
very early in the spring. It is a Plum and it blows at the same
time, or a very little earlier, than the Plums. It is a remarkable
fact that there is always, that is every year of our lives, a spell
of cold and angry weather just at the time that this hardy little
tree is in bloom. The country people call it the *Black Thorn
winter* and thus it has been called, I dare say by all the in-
habitants of this island, from generation to generation, for a
thousand years.

This thorn is as hardy as the White Thorn; its thorns are
sharper and longer; it grows as fast; its wood is a great deal
harder and more tough; it throws out a great deal more in
side-shoots; and it is, in every respect, better than the Hawthorn for the making of a Hedge. If I be asked how it has
happened, then, that the Hawthorn is constantly used for this
purpose, and the Black Thorn never, or scarcely ever, I
answer that the reason is very clear; namely that a sack of the
seed of a Hawthorn may, almost anywhere, be got for a shilling
or half a crown at the most; and that, to get a number of Black
Thorn sloes, equal in number to the Hawthorn berries contained in a sack, would, in almost any part of the kingdom, cost
five, ten, nay twenty pounds.

The sloe is very large compared with the size of the Hawthorn berry; you must get six sacks perhaps of the sloes to have
a number equal to the berries contained in one sack; and six
sacks of sloes, except in very woody countries, would not be
found perhaps in the half of a whole county. The tree, like
other plums, is liable to blight. It seldom bears any considerable
crop, and very frequently bears no fruit at all. It grows no
where except in hedge-rows and coppices; in the former it is
too much exposed to bear much fruit; and in the latter, it is
too much in the shade to bear any fruit at all. Hence it is, that,
though all of us who have been born and bred in the country
know that the Blackthorn is by far the best of the two, we
never heard of such a thing as the planting of a Black Thorn
Hedge.

As *bushes*, for the making of hedges, the Black Thorn is always carefully laid by when hedge-rows and coppices are cut. These bushes will lay longer in a dead hedge without perishing than any other sort of stuff of which dead hedges are made. This Thorn will thrive, and that vigorously too, in the very poorest of land. It sends up straighter shoots from the stem than the White Thorn does, and these shoots send out, from their very bottom, numerous and vigorous side-shoots, all armed with sharp thorns. The knots produced by these side-shoots are so thickly set, that, when the shoot is cut, whether it be little or big, it makes the most beautiful of all walking or riding sticks. The bark, which is precisely of the colour of the Horse Chestnut fruit and as smooth and as bright, needs no polish; and, ornamented by the numerous knots, the stick is the very prettiest that can be conceived. Little do the bucks, when they are drinking Port wine (good old rough Port) imagine that, by possibility, the beautiful stick with which they are tapping the sole of their boot, while admiring their legs, never does their philosophy carry them so far as to lead them to reflect that, by possibility, for the 'fine old Port,' which has caused them so much pleasure, they are indebted to the very stick with which they are caressing their admired Wellington boots.

In some situations, it would not be difficult to obtain Black Thorn seed enough to plant a hedge of considerable length; and, at any rate, it may be done if any one will take the pains; and, therefore, I shall now proceed to state the manner of raising the plants. The seeds are not properly ripe until pretty late in the month of October. They may be suffered to hang till they are dead ripe, provided the boys do not find them out; for though, as we have seen, excellent in the making of Port wine, they are rather too astringent, too 'rough' for the birds, whose tastes seem to differ from the lovers of *Good Old Port*. When ripe, they should be gathered; mixed with dampish sand, kept turned in a cellar or a shed, until the month of February, and then sowed in beds.

Like the seed of the plum, that of the Black Thorn comes

up the first year; that is to say, if not sowed too late, and if kept in moist sand or earth until the time of sowing; and without these precautions, Plum-stones will lie a whole year before they begin to sprout, as Peach and Apricot stones will. If managed in the manner that I have here directed, the Black Thorn plants will be up in the month of May, and in the month of October afterwards, they will be from five to eight inches high. In the month of November or next in the month of March, they ought to be removed into a nursery, being assorted and planted.

As an ornamental shrub, or little tree, the Black Thorn is by no means equal in beauty to the Hawthorn; neither the shape nor colour of the leaf is equal to that of the White Thorn nor is the leaf nearly so abundant; in proportion to the size of the tree. But, the Black Thorn comes into full bloom a month if not six weeks before the Hawthorn; and it makes a very gay show, when scarcely any blossoms have appeared, or any tree is in leaf. Clumps of Black Thorns, therefore, or independent plants of this kind, might be placed very advantageously in parks and lawns; and if managed well, even in shrubberies, for they are in bloom much earlier than any other shrub. The plant has so many advantages over the Hawthorn, that it is impossible that it should not be cultivated, in so many cases in preference to the Hawthorn, were it not for the great difficulty of obtaining the seed in any considerable quantity.

The Yew. 'Though last not least', to use a saying that has been more than worn out for these last five hundred years. This is our native English Cedar. Its outward appearances are well known to us all; for, first or last, the most of us have seen a church-yard, however few, comparatively, may have been within the church, and there is scarcely a church-yard in England that does not present a Yew tree to the eyes of the beholder. The Yew tree sometimes rises to the height of forty or fifty feet, and would go a great deal higher if attention were paid in the pruning the side-shoots as the tree increases in height. It is frequently a very large tree. I have seen several, each of which has been more than fifteen feet round the trunk; and as I mentioned in the Rural Ride performed in the month of August, 1823, I measured the Yew tree in the church-yard of the village of Selborne, in Hampshire, which was, at some distance from the ground, *twenty-three feet and eight inches in circumference.* I had not the means of making a very exact measurement; but, my error could not be very great; and this is a monstrous circumference.

The Yew tree attains an age beyond the possibility of human ascertainment, unless some national record were kept of the matter. Its growth is slow; but it appears never to lose any part of that which it gains. The wood surpasses, very far indeed, all other English wood, in point of durability, and of strength, and elasticity joined to both these. It has a red heart, like the imperishable Red Cedar of America; but it is not odoriferous like the wood of the Red Cedar: and it is very heavy, while the Red Cedar is very light. The wood never perishes, or, at least, so say the country people. Gate-posts are sometimes made of it, and they have never been known to rot. It is very fine-grained and receives a very high polish. I have seen a kitchen dresser made of it, which being kept in good order, was as shining as any piece of furniture I ever saw. It is super-excellent for making the bows of the backs of wooden chairs; for making ox-bows, and, it is said, the bows, used by warriors formerly in this country, were made of this wood. Even the slender shoots of it are as tough, or tougher, than those of the Hickory; and when they happen to get any considerable length, they become whip-handles, and other things, where great toughness and elasticity are required.

The Yew appears to grow pretty nearly equally well upon all sorts of land, shallow or deep, dry or wet; but it seems, from the frequency of our finding it on such land, to have been formerly, at any rate, grown principally upon chalky land. It resists all weather, stands uninjured on the bleakest of hills, where even the scrubbiest of thorns and underwood will hardly live. Big as the head of this tree generally is in proportion to its trunk, most heavily laden as it constantly is with leaf; forming as it does, such a hold for the wind, neither head nor trunk ever flinches, though in situations where it would be impossible to make an Oak grow, and where no other large tree could be prevented from being blown out of the ground.

The Yew is, or rather has been (for it is not much the fashion now), used for making *hedges*, as screens in and near gardens and round about houses, for which purpose it is the best material that can possibly be conceived. It is easily clipped into any form; its twigs are delicate and tough; it can be shaved close down to the ground; and it makes a fence or screen through which no wind can find its way; and being *ever-green*, it is very valuable on this account. People had more patience and diligence formerly than they have now; and, therefore, I have never heard of the planting of a Yew hedge in the whole course of my life. There are many, however, still remaining in England, but these have all descended to us from our ancestors, who lived in those dark ages, when men were foolish enough to think that patience ought to be required in the acquisition of things of great value, and that large estates ought not to be acquired in a few years by merely 'watching the turn of the market'. There is a hedge of this sort at *Petworth* in *Sussex*, which, if I recollect rightly, I judged to be eleven feet wide, kept very closely and neatly clipped all across the top, and on the sides, from the top to the bottom. The bottom on one side of the hedge meets the top of a wall which supports a bank against a road or street. I was quite as much delighted with this object, as I was with the house and park of Lord Egremont, the gate that opens to which, is not more than two hundred yards distance from the hedge.

PESTS

Ants – A very pretty subject for poets, but a most dismal one for gardeners; for it is one of the most mischievous of all things, and most difficult of all to guard against or to destroy. It is mischievous in many ways, and all the sorts of ants are equally mischievous. Those which have their nests in little hillocks on the ground; that is to say, the small ant, is the sort which most frequently display their mischievous industry in the gardens. I once had a melon-bed that underwent a regular attack from the community of *horse-ants*, as the country people call them; that is, the largest ant that we know anything of. I know nothing but fire or boiling water, or squeezing to death, that will destroy ants; and, if you pour boiling water on their nests in the grass, you destroy the grass; set fire to a nest of the great ants, and you burn up the hedge or the trees, or whatever else is in the neighbourhood. As to squeezing them to death, they are amongst the twigs and roots of your trees and plants; they are in the blossoms, and creeping all about the fruit; so that, to destroy them in this way, you must destroy that also which you wish to protect against their depredations. Ants injure everything that they touch; but they are particularly mischievous with regard to wall-trees: where they attack successively bud, blossom, leaf, and fruit. There is no method of keeping them from the wall. They may be kept from mounting espaliers by putting tar round the stem of the tree, and round the stakes that the limbs are tied to; but there is no keeping them from the wall, unless by killing them. Mr Forsyth recommended to make the ground very smooth near the bottom of the tree that they attacked; then to make smooth holes with a sharp-pointed stake or iron bar, down into which, he says, they will go; and then he recommends to pour water into these holes and drown them. Monsieur de Comble recommends the laying of sheep's trotters or cow-heels with the skin on, near the attacked tree, and that, when these be well covered with ants, to plunge them into a bucket of water, drown the ants, then put the sheep's trotters near the tree again to wait for another cargo. By these means something may be done, to be sure; but, the true way is, to find out the nest from which they come; for they are extremely scrupulous in this respect; it is only one tribe that makes its attack upon one and the same object. If

you look attentively, you will find that, in the morning, very
early, they all come in the same direction, and that they go in
exactly the same way back at night. Trace them to their for-
tress; and, when it is quite night, treat them to a bucket of
water that is as nearly upon the boil as possible. You kill the
whole tribe. When my melon-beds were attacked by the
horse-ants, I set to work to discover whence they came. I
traced them along a brick wall. Then out of the garden between the door-frame and the wall. Then along at the bottom
of the edge of the wall on the side of a lawn; then, after having
made an angle along the wall, going, as I thought, over it into
a meadow on the other side. Every corner of hedge and ditch
of that meadow was examined to discover the nest, but in vain.
Looking back to the spot where I thought they went over the
wall, we discovered that they turned along the top of the wall,
and went under the roof of a summer-house that was ceiled
below: having lifted up a tile, there we saw bushels of ants with
little sticks and straws, the result of years of their detestable
industry. A copper of water was made to boil against the
evening. It was taken to the spot in a boiling state as nearly as
possible; everything was prepared for the purpose, and by mid-
night, scarcely a handful of them were left alive; and my melon-
bed, which I was actually upon the point of giving up as lost,
was suffered to proceed unmolested. The greatest care, therefore, ought to be taken, especially if grass ground be near the
garden, to hunt out ants' nests, and to destroy them.
Birds – The way to keep birds from fruit, and, indeed, from
everything else, is to shoot them, or frighten them away, or
cover over effectually with nets the object which they covet.
I have spoken occasionally of the care to be taken in this
respect; but, in all cases, where birds are very fond of the
thing you have, you must keep them away, or give up the
cultivation of the thing; for it is time and labour thrown away,
to raise things and then let them be destroyed in this manner.
There is one season when to defend yourself is very difficult;
I mean the spring, when the birds attack the *buds*. There are

certain buds which the sparrows will destroy, just when they
are sending out their fruit: but the great enemies of buds are
the bullfinches, the chaffinches, and, above all, the greenfinches,
which assail the buds of plums of all sorts in a most furious
manner. They are hard driven for food at this time of the year;
and they will actually strip whole branches. It is, however,
contended by some persons, that, after all, they do no harm;
for that, there are insects in the bud which they eat; and that it
is not the herbage that they want, but the animal, seeing that
birds live upon grain, and pulse, and insects, and not upon
green things. This is by no means true: they do live upon
green things, or at least they eat them, as we see fowls eating
grass, during a great part of every day. I believe that these little
birds eat the buds, and are not at all looking after insects. The
wild pigeons in America live, for about a month, entirely upon
the buds of the sugar maple, and are killed by hundreds of
thousands, by persons who erect bough-houses, and remain in
a maple wood with guns and powder and shot, for that purpose.
If we open the craw of one of these little birds, we find in it
green stuff of various descriptions, and, generally, more or less
of grass, and therefore it is a little too much to believe, that, in
taking away our buds, they merely relieve us from the insects
that would, in time, eat us up. To keep birds from buds is a
difficult matter. You cannot net all your trees; nor can you
fire with shot among your trees without doing a greater harm
than that which you wish to prevent. Birds are exceedingly
cunning in their generation; but, luckily for us gardeners, they
do not know how to distinguish betweeen the report of a gun
loaded with powder and shot, and one that is only loaded with
powder. Very frequent firing with powder will alarm them so
that they will quit the spot, or, at least, be so timid as to become
comparatively little mischievous; but there is what, to me, is a
recent discovery in this matter, and which I have hitherto
practised with complete effect. It has the great recommendation
of good inventions, perfect simplicity: having a bed of radishes
or other things that you wish to keep birds from coming upon,

stick a parcel of little pegs about a foot long into the sides of the bed, at distances of about three yards apart, and then take a ball of coarse white sewing cotton, tie the end of it to the top of one of these little sticks, and then strain the cotton on to another, fastening it round the top of every stick, and going in a zig-zag across the bed. What the little picking and scratching devils *think* of these threads I know not, but it keeps them off, and that is enough for our purpose. I imagine that it inspires them with *doubt*, and as doubt has great influence upon the human race, why should it not have the same upon these timid and watchful creatures?

Black Grub. It should be called the brown grub, for it is not black. In its workings, it is half way between a rook-worm and a caterpillar. It lies snugly under the ground near the roots of the plant in the day-time, and comes up at night, eats the plant off at the stem, or eats out its heart. This is a most perverse as well as a most pernicious thing; it is not content, like the caterpillar, the snail or the slug, to feed upon the leaves; but it must needs bite out the heart, or just cut off the plant at the bottom. Lime has no power over it: nothing will keep it off: no means but taking it by the hand: in a garden this may be done, by examining a little about the ground just round the stem of every plant; for as soon as it has destroyed one plant, it gets ready for another for the next night's work. In a garden this thing may be destroyed or kept down; but in a field it is impossible and many a field has had its crop almost totally destroyed by this grub.

Caterpillar. Very few more mischievous creatures than this infest the gardens. In the first place, it is a most destructive enemy of fruit-trees; apples, pears, plums, quinces, medlars and gooseberries, but particularly apples and plums are literally flayed alive by this nasty insect. Hundreds of trees together are, early in the month of June, very frequently completely stripped of every leaf by the caterpillars. Of their progenitors I know little; but that they appear in the winter, when the leaf has fallen, as a little crusty shell-like ring fastened tightly round the twigs of the tree, and generally upon apple-trees. This crust is not more than half an inch long, and it is pricked all over in regular rows of holes, looking something like a piece of an old thimble twisted round the twig. In the spring a swarm of little caterpillars issues from this crust, and works its way all over the tree, and, to an ordinary observer they make their first appearance in a web formed into the shape of a bag or sort of wallet attached to the branches of trees. And this bag is a small thing at first; but it grows larger and larger as the caterpillars within it increase in size. If you open one of these bags, a goodly tribe glads your sight; and, if you leave the bag till the caterpillars grow too big for it and open it themselves, they sally forth in every direction, and strip the tree of its leaves. Prevention is not, however, in this case, very difficult. If they come on espaliers, you pick the bag off as soon as you perceive it, and crush it under your foot. If they come on standard-trees, you must take a ladder; but a better way is, to load a gun with powder, and blow the bags from the trees. If once they escape from the bag and go on their travels, you have no remedy. If you shake the tree and bring part of them to the ground, they crawl up again. Lime has no effect upon them; and your only hope is that your other enemies, the sparrows, will lend their assistance in delivering you from these; and I do verily believe, that, were it not for the sparrows, and other birds, these insects would make it next to impossible to cultivate gardens in England. They have no slugs and snails in America; but caterpillars they have, and they sometimes strip an orchard of every one of its leaves. There are caterpillars which infest the cabbages and the Swedish turnip, and some other herbaceous plants. These manifestly proceed from the butterfly; but, unfortunately, they do not make their appearance in little pockets or bags; but you make the first discovery of the honour of the visit that they are paying you by perceiving their gnawings upon the edgings of the leaves of the plants. Let them alone for a little while and they will go from cabbage to cabbage until there is not a bit of leaf left in the whole patch.

They leave you the skeleton of a cabbage, taking away all the flesh, and leaving all the bones; that is to say, the stalk of the cabbage and the ribs of the leaves. These are most mischievous things; they are wholly insensible to the powers of lime: in heat they delight: wet will not injure them; frost is their only destroyer; and many a time have I prayed for winter in order to see an end of the caterpillars. In order to mitigate the mischief, and, indeed, in a great measure to put a stop to it, look narrowly among your plants of the cabbage kind about the middle of the summer. If you see the butterflies busy, expect their followers in due time. Watch the plants: as soon as you see one attacked take it entirely up, shake the caterpillars from it upon the ground, put them to death with your foot, and carry the plant away to the pigs. 'Tis very rarely that the whole or any considerable part of a piece of cabbages is attacked at once; and therefore you may, in some measure, guard against the mischiefs of this pernicious insect, of which there are several sorts, some green, some brown, some smooth, some hairy, and all equally mischievous.

Ear-wig. This is a most pernicious insect, which feeds on flowers and on fruit, and which, if it congregated like the ant, would actually destroy everything of this sort. Its favourite flowers are those of the carnation kind. To protect very curious plants against them, the florists put their stages on legs, and surround each leg with a circle of water contained in a dish which is so constructed as to admit the leg through the middle of it, seeing that the ear-wig is no swimmer. Others make little things of paper like extinguishers and put them on the tops of the sticks to which the carnation-stalks are tied. The ear-wigs commit their depredations in the night, and they find these extinguishers most delightful retreats from the angry eye of man and from the burning rays of the sun. Take off the extinguishers, however, in the morning, give them a rap over a basin of water and the enjoyments of the ear-wigs are put an end to at once. They are very nasty things in fruit of the stone kind, and particularly the apricot. They make a way in the foot-stalk of the fruit, get to the stone and live there day and night; so that, when you open a fine apricot, you frequently find its fine juice half-poisoned by three or four of these nasty insects. As soon, therefore, as the wall-fruit begins to change its colour, the tree should be well furnished with extinguishers made of cartridge-paper and able to resist a shower. By great attention in this way you destroy them before the fruit be ripe enough for them to enter. But, one great protection against all these creeping things is to stir the ground very frequently along the foot of the wall. That is their great place of resort; and frequent stirring and making the ground very fine, disturbs the peace of their numerous families, gives them trouble, makes them uneasy and finally harasses them to death.

Flies. Great flies, like the flesh-flies, feed upon all the softer fruits; and even upon apples and pears. They are destroyed or kept down precisely in the manner directed for the wasps. Some persons, in order to preserve fine pears, cover them over with bunting, a piece of which they tie completely over each pear: this is a very troublesome, but a very effectual method.

Mice. Very troublesome creatures. They commit their depredations by night and must be well looked after. Brick traps are the best things; for, as to poisoning them, you may poison at the same time your cat or your dog. Great vigilance, however, is required to keep down mice; but it ought to be resolutely done.

Moles. These cannot get into a garden with a wall round it. If they come through or under the hedge, and make their workings visible, they ought to be caught without delay; for, if suffered to get to a head, they do a great deal of mischief, besides the ugliness which they produce.

Rats. If the garden be near to a house or outbuildings, and especially near to a farm-yard, where dogs and ferrets are not pretty constantly in motion, the rats will be large sharers in the finest of the fruit that the garden produces. On the walls, in the melon-bed, even in the strawberry-beds, they will take away the prime of the dessert. They do but taste, indeed, of

each, but then they are guests that one does not like to eat with. Here is absolutely no remedy other than dogs and ferrets. I have seen a wall of grapes pretty nearly cleared by rats, some farm buildings being at the backside of the wall: these nasty things must, therefore, be destroyed by one means or another. *Rook-worm.* This is an underground enemy; a miner and sapper. It is a short worm or long maggot as big round as a thick goose-quill, body white, and head partly red and partly black. It is a fact, I suppose, that the May bug, or chaffer, comes from this worm. The French call it the *ver hanneton* which corroborates that opinion. It attacks the roots of plants, and will even attack the roots of trees, and will now-and-then destroy some young trees. It will clear a patch of cabbages in a very short time. It is under-ground, and therefore not to be guarded against; but a garden may very soon be ridded of it. First, kill every one that you meet with in digging; next, the moment you see a plant begin to flag, dig it up and take up the worm. If the worm be on its travels, you are sure that it is gone towards the next adjoining plant to the right or to the left. Pursue it both ways with the spade, and ten to one but you overtake it. A little perseverance in this way will soon clear a garden of the rook-worm; but as to our fields, their crops would be absolutely devoured, in many cases; or, rather, the plants would be destroyed, were it not for the rooks, which are amongst the most useful of the animals in this country; and really it is too hard to grudge them a little of the corn when they have so largely contributed towards bringing the whole of it to perfection.

Slug. This is a snail without a shell, and like the snail likes neither sun nor frost. Some slugs are black, others whitish, others yellow. The great black slug and the yellow slug live chiefly upon worms, and do not touch plants of any kind. The mischievous thing is the little slugs that hide themselves in the ground or under grass or leaves, and that come out at night, or in the rain, and eat the garden plants of almost every description more or less, and sometimes pretty nearly clear a field of wheat. Slugs cannot live under the shining sun, nor can they move about much except when the ground is wet or moist from dew or rain; then it is that they come forth and make up for lost time. They are propagated amongst weeds and grass, and anything that affords constant shade and tranquillity. A garden constantly clean is, therefore, the most effectual prevention; but if they come, they must absolutely be killed, or you must give up your crop. The way to kill them is this. Take *hot lime*, in a powdered state, put it into a coarsish bag; and after nightfall or before sun-rise, in the dew, or on the moist ground, go over their haunts, shake the bag and let the fine powder fall upon the ground: some little particle will fall upon every slug that is abroad; and every slug that is touched with the lime will die. If rain come it will destroy the power of the lime, and then it will be necessary perhaps, for you to repeat the remedy several different times.

Snails. From the curious construction of the snail, it is known to everybody in town as well as country. It is very mischievous, and especially amongst fruit-trees, where it annoys the fruit, as well as the leaf, but particularly the fruit. It is a great enemy of the apricot and the plum, both of which it will eat whether in the green or in the ripe state. It is very mischievous amongst the plants in the garden in general; but its size and its habits and manners make it not difficult to destroy. Its places of harbour are behind the trunks or big limbs of wall-trees, in a garden, or, round the butts of the trees that form the hedge of the outside of the garden. Snails lie in such places all the winter long, and never stir till they are warmed into life in the spring. Many persons have kept snails for a year or more nailed up in a box and have found them just as lively afterwards as if they had never fasted at all. In winter-time, in dry and frosty weather, snails should be routed out from all their fastnesses, and destroyed. This is the most effectual way of guarding against their depredations; for, when the leaves come out, they have shelter, they are exceedingly cunning in availing themselves of that shelter, but though you finally discover and kill them, they spoil your fruit first.

Spider. I do not know that the common spider does any harm to the gardener, and I know that it frequently does good by killing the flies, but there is a red spider which is very mischievous to vines, especially when under glass. If attended to, however, they are easily destroyed, and the destruction of them should not be neglected. Plentiful washing of the trees with water is the great remedy, and, in hot-houses, syringes are made use of for this purpose.

Wasps. These are enemies of another sort, and, in some years, most troublesome they are. They fix upon the finest fruit, and in some seasons, long before it be ripe. They will eat a greengage plum to a shell; and, while they spoil your fruit, they will not scruple to sting you if you come to interrupt their enjoyment. The first thing to do is to destroy all the wasps' nests that you can find anywhere in the neighbourhood. These nests are generally in banks. Discover the nest in the day-time, open it with a spade at night and pour in boiling water. There is a little bird called the red-start that destroys the wasps; but boys are their great enemies: and about sixpence a nest will keep any neighbourhood pretty clear of wasps. But the great remedy is to kill them when they come to the tree, and that is done in this way; you fill a pretty large phial half full of beer mixed with brown sugar; the wasps attracted by this go down into the phial and never come out again. The phials must be emptied every day, if anything like full, and put up again with fresh sugar and beer. A string is tied round the neck of the phial, which is thus fastened round some part of the tree. There must, however, be a considerable number of these phials attached to every tree.

Wire-Worm. This is a little yellow worm which, at full growth, is about an inch long; and is called *wire-worm* because it is very tough and difficult to pinch asunder. It is bred in grass-land, and in old tufts of grass in arable land. A piece of land digged or ploughed up from a meadow, or grass-field, will, for a year or two, be full of these worms, which carry off whole fields of wheat sometimes. In gardens they are very destructive. They attack tender-rooted plants, make a hole on one side of the tap-root, and work their way upwards till they come to the heart. You perceive when they are at work, by the plant dropping its leaves; and the only remedy is to watch the plants narrowly and, as soon as you perceive the tips of the leaves beginning to flag, to take it up and destroy the worm. They are particularly fond of lettuces that have been transplanted; and I have had whole rows of lettuces destroyed by these worms, in spite of every precaution.

Wood-Louse. It is a little grey-coloured insect of a flat shape and about twice as long as it is broad. When you touch it, or when it sees itself in danger, it forms itself into a ball, and very much resembles Dutch cheese, and is, by the children in the country, called a *cheese-bob*. Its name of *wood-louse* comes from its habit of living and breeding in rotten wood and under boards or slabs that are lying upon the ground; but it also haunts very much the cracks in bricks, and the holes in the joints of walls. It feeds upon buds and blossoms, and also upon the fruit itself. When it gets into hot-beds, it hides round the edge of the frame, and does a great deal of mischief to the plants, especially when they are young. Cabbage-leaves or lettuce-leaves laid in a hot-bed or against the edge of the wall, will invite them to take shelter as a place of retreat for the day, all the dilapidations being committed in the night. You lift the leaves in the day-time and kill them; and, further, as to walls, the great remedy is to keep all the joints well pointed, and to fill up any cracks that there may be in the bricks.

GARDENERS

Every gardener thinks that every one who employs him is, as far as relates to gardening, a natural-born fool. He will allow him to be, and indeed he will boast of his being, the greatest of orators, the greatest of generals, the most valiant of admirals, the most profoundly wise of lawgivers, the most heavenly of all heaven-born ministers, the most pious and learned of bishops, the most learned of all learned lawyers, and, of a physician, capable, almost, of raising the dead to life; but that, in matters of *gardening*, he will insist that he is essentially a fool, and that he does not know, and ought not to know, any more about the raising of a tree than he, the gardener, knows of any of the learned professions, a profound knowledge in which he is already (with and without cause) to ascribe to his master. One of the consequences of this way of thinking is, that gardeners, if the master be of a character that makes it perilous to flatly contradict him, hear, with very little interruption, all that he has to say, and all that he relates to them as having been said by others. They receive his directions very quietly, then go away, and pay no more attention to them than to the whist-ling of the winds; and as to *books* that may be put into their hands, if not written by a professed gardener by trade, they would laugh at the idea of any one supposing it possible that they can contain anything worth looking at.

The best, and, in the end, the *cheapest* way is to employ men by the day; to have a really good and trusty man to *work with them* (example is better than scolding); to see them *begin well* yourself; to visit them often; to repeat, at *every visit* (for their memories are short), your orders as to the manner of doing the work, and to insist on their keeping steadily at work, for if men *keep on,* they will almost always do work enough. The straight back and the gossip are the great enemies of the progress of the labours of the field. But the great things of all (next after sufficient pay), are your *own presence* on, or near, the spot, and a conviction in the minds of the men, that *you understand the whole of the business well.* If you *could* just take the spade, or spud, and *show* them a little now-and-then; if you *could* do such a thing, it would be a great additional benefit; and I pledge my word to you, that it would do harm to neither your body nor your mind. The two first fingers of my right hand are still somewhat bent, from having been, more than forty years ago, so often in close embrace with the eye of the spade and the handle of the hoe, but I do not find that this bend makes them the less fit for the use to which they are at present applied. What think you of a short lesson in the garden every day before breakfast for a month, before your trenching begin? This would be attended with one signal and most important advantage; namely, it would, by the appetite it would give you, enable you to judge what portion of food ought to be allotted to the man who lifts six thousand heavy spits a day. And, as to the character of such an act, the Emperor of China holds the plough once a year; and, besides, why should it be more degrading to you to be as skilful as your gardener, than it is to be as skilful in the infinitely lower business of your groom, or your dog-feeder? However, if this manual act be too much to expect, let me exhort you to let your workmen be thoroughly convinced that *you understand all the matter well.* Let me exhort you to give simple and positive *orders,* and never, no never, to encourage, by your hesitation, even your bailiff or gardener so much as to *offer* you *advice.*

A KALENDAR
of Work to be performed in each Month of the Year.

I went by the field of the slothful, and by the vineyard of the man void of understanding and, lo! it was all grown over with thorns, and nettles covered the face thereof, and the stone wall thereof was broken down. Then I saw and considered it well: I looked upon it, and received instruction.

Proverbs XXIV. 30

JANUARY

KITCHEN GARDEN – In our variable climate, what is to be done this month, depends much on the state of the weather; but, if it is not deep snow, there is always something to be done advantageously. Even deep snow gives time for cleaning, thrashing, and sorting of seeds, preparing stakes and pea-sticks, tying mats, sorting bulbs, and many similar sorts of employment. Dry frost makes an opportunity of manuring land with ease and neatness, and also of pruning gooseberries, currants, and other hardy shrubs, and of clearing away dead trees and bushes, and thinning others. If the weather is mild and open, a few seeds may be sown, but not in great quantities; in general, land is now troublesome to work; fresh digged, or forked (if it has been ridge-trenched), works best. Sow, for early use, radishes and carrots on a warm sunny border; peas, beans, round spinage, parsley; small salad in frames, old mint roots on heat under glass will soon give a supply of green mint. Pot over and cover seacale, and rhubarb; where a succession is required, some may be forced on heat, some with dry litter, ashes, or even with light mould. Remember that the blanching of seacale depends on the exclusion of the light and air, either of which spoils it. Attend to lettuces and cauliflowers in frames, endive in frames or under hoops; in mild weather, if dry, let them have as much air as possible; in hard frost, cover well. A few onions for eating green should now be sown, cover with dry litter till they are up. Mushroom beds must be most carefully protected from wet and frost, cover well with dry litter; mats or canvass covering is indispensably necessary, over the litter, to keep it firm and throw off the wet. Turn dunghills and compost heaps. I say it at once for the whole year, destroy vermin wherever you can find them.

FRUIT GARDEN – Cut and nail in vines, pears, cherries, and plums, against walls and fences; apricots and peaches are more safely left till next month, or rather till the buds *begin* to swell. Thin standard and dwarf trees of decayed and superfluous wood. *Plant* fruit trees if favourable weather; prune gooseberries and currants and other fruit shrubs; prepare and plant cuttings of the two former, and suckers of the latter, to plant in a nursery; take off the shoot buds of the gooseberries and currants as far as they are to be *planted* in the ground, the so doing prevents the stems of the future bushes from being always pestered with suckers; it will not prevent the cutting from striking.

FLOWER GARDEN AND SHRUBBERY – In this, little can be done this month, except planting hardy roots and bulbs; protecting by covering those planted in autumn; *plant* deciduous trees and shrubs. *Sow* sweet-peas to succeed those sown in November; roll, poll, and sweep grass walks, verges, and lawns; keep them clear from leaves and rubbish; root out dandelions and docks if there. Plant box edgings.

FORCING GARDEN – Attend to articles sown on heat in December. If you want them early, you may now force radishes on slight heat under frames, asparagus, potatoes, French beans, rhubarb, seacale, strawberries *in pots*. Where you have the convenience of a green-house, many things may be forwarded in this way. Cucumbers and melons now require particular attention, to keep up the heat equally by linings and coverings. *Sow* for next month.

GREEN-HOUSE – Pick off decayed leaves, water sparingly, give air freely in mild *dry* days; light fires, not to promote growth, but to keep out frost, and to dry up damps. Cover potted stocks and other plants in frames during hard frost.

FEBRUARY

KITCHEN GARDEN – Whatever was mentioned as proper to be

done last month, and was not performed, may be done this, either in sowing, planting or pruning: if your land is light and dry, you may sow onions for a main crop towards the close of the month, but on cold and heavy soils it is better to wait to the middle of March; earth up celery, if you have any left unearthed. *Plant* beans for a full crop about the middle of the month. Sow peas in succession as they are likely to be wanted, radishes under straw covering, coss lettuce *on heat to transplant;* on light rich soils lettuces may be sown broadcast, or drilled; they will, if they escape slugs and other vermin, be nearly as forward as those now sown on heat; the frost will seldom injure them. If you have good strong plants of the coss lettuces which, sown in October, in the frames, have stood the winter, and the weather be mild and the plants in a growing state, transplant them on a sunny spot of rich light soil; do not be afraid of winds and frosts. *Sow* savoys for early planting, leeks, cabbages if wanted, spinach, parsnips, parsley, carrots, Dutch turnips under litter, cauliflowers on heat. Many of these sowings must depend on the nature of your soil and the weather. *Plant* cabbages, garllck, rocambolo, onions for seed, shalots, cives, horseradish-crowns. Get what hoeing you can done in dry days; get as close in with your work as you can; and prepare for the busy month of March.

FRUIT GARDEN – Continue, and conclude if you can, the business of last month; prune and nail in peaches, apricots, figs, nectarines; lay vines (in pots is the best way), *sow* haws, pips and fruit kernels; many kinds of which may be not only much forwarded, but *secured*, if sown under glass, from the birds and vermin; in their infant state many of these things are very tender. Plant fruit-trees, shrubs, and bushes; crab, apple, pear, and other *stocks* for next year's grafting; if favourable weather, grafting may be begun at the close of the month; collect and prepare scions for grafting.

FLOWER GARDEN – Towards the end, if open weather, transplant pinks, carnations, and other hardy flowers, part the roots of southernwood, sweet-williams, candytuft, campanula etc

if not done in the autumn; make and repair box and thrift
edging. Get your strawberries, grass lawns and verges and
gravel walks, into order.

FORCING GARDEN – Attend to your cucumbers and melons,
and to all other matters in this department. *Sow* cucumbers
and melons to pot and ridge out in March; many articles may
now be forced with less difficulty than in the former month,
as, kidney-beans, strawberries, rhubarb; less heat and less
covering will be required unless the weather is particularly
severe.

GREEN-HOUSE – Give air freely when the weather admits of so
doing, no more fire-heat than is necessary to keep out frost
and dispel damp; other management as last month; shorten
and head straggling growing plants.

MARCH

KITCHEN GARDEN – Sow artichokes, Windsor beans, cauliflowers
to come in the autumn; celery, capsicums, love-apples,
marjorum and basil, on gentle heat; lettuces, marigold, blue,
Prussian and other peas in succession; onions for a principal
crop, but do not sow them till the ground works well and fine;
parsley, radishes, borage, savoys, small salading in succession
as wanted; asparagus in seed beds, beets, salsafy, scorzenera,
skirrets, fennel, cabbages red and white, turnips, nasturtiums,
early purple brocoli, thyme, and all sorts of herbs that are raised
from seed, Brussels sprouts, parsnips, round spinage, leeks,
carrots for a main crop, chervil, coriander, French beans at the
close of the month in a warm soil; *plant* out cauliflowers, hops
in clumps for their tops, small onions of last year's sowing,
for early heading, old onions for seed, lettuces, perennial
herbs by slips or parting the roots, asparagus, artichokes from
suckers, potatoes for a main crop, cabbages, white and red,
Jerusalem artichokes, chives, potato-onions, etc.

FRUIT GARDEN – Head down old trees and shrubs, plant out
trees and shrubs, continue and *finish* pruning if you can; plant
out stocks for next year's grafting and budding, if not done in

autumn, this is the principal month for grafting most sorts of trees, plant gooseberries, currants, raspberries, protect blossoms of fruit-trees when in blossom, dig fruit-tree borders.

FLOWER GARDEN – *Sow* adonis, alyssum, prince's feather, snap-dragon, yellow balsam, candy tuft, catchfly, convolvulus minor, devil-in-a-bush, hawkweed, Indian pink, larkspurs, lavatera, linaria, mignonette, moonwort, nasturtiums, nigella, palma christi, pansey, sweet pea, persicaria, scabious, sunflowers, strawberry-spinage, ten-week-stocks, sweet sultan, venus-navel-wort. On *hot-bed*, *sow* convolvulus major, amaranthus, tricolor and globe, balsam, chinaster, china hollyhock, chrysanthemums, jacobea (French groundsel), ten-week-stocks, zinnia, marvel-of-Peru, plant autumnal bulbs, such as the tiger flower, dahlias, anemones, and ranunculuses, if any not yet planted, may make a late blossom. Hardy kinds of potted plants that have been sheltered should be now gradually inured to the open air, dress auriculas, carnations, protect best tulips and hyacinths, plant offsets, and part fibrous-rooted plants, take up and plant layers of carnations, pinks, seedlings of the same and other things and plant them, plant box, thrift, and daisies for edgings. Many of even the more hardy kinds of plants will be much advanced, if sown on a little heat, and carefully hardened to the open air before put out. Lay turf, put and *keep* gravel-walks in good order, also grass-plats and edgings, roll, poll, and sweep, keep the shrubbery clean, remove all litter, finish planting shrubs.

FORCING GARDEN – This is a good time to begin to force vines, if you have a *grapery,* the sun will so materially assist you. Cucumbers and melons, in fruit, and beginning to show fruit, must be carefully attended to; give air and water as required; line the beds as the heat declines; cover carefully, and uncover early; sow for a crop for June and August; make slight hot beds for French beans, and for raising seedlings of the tender plants you may want.

Mushroom beds may be made.

GREEN-HOUSE – Fire heat will not be necessary now, unless

in unusually cold or damp weather; or, when frost is indicated, give air and water more freely. Shift and re-pot those plants that require it. Propagate by cuttings, slips, layers, and parting the roots, grafting, budding, and inarching. Sow geranium and balm-of-Gilead seeds, etc, on slight heat. Pot out in small pots when fit, the balsams, amaranthuses, cockscombs, sensitives, and other tender annuals, on slight heat under glass; they will, the best of them, have to take their station in the green-house, when the present possessors, the geraniums, myrtles, etc, are turned out. *See Flower Garden, March.*

APRIL

KITCHEN GARDEN – Beans may still be sown, peas, kidney-beans, scarlets, beets, brocoli, purple, white, and brimstone, late and early of all sorts; the cape about the third week, the sprouting the first week, savoys, cabbages, green cale, brown cale, Lapland ditto, and sea cale, it is best to sow this last in drills where it is to remain, as it transplants with great difficulty and never well, on account of the brittleness of the roots; sow it in drills about eighteen inches apart, from row to row. Continue sowing in general what was directed last month if wanted, and not then sown, or the seed or plants destroyed or failed. Kidney-beans for full crop at the end of the month if dry, also scarlet-runners, herbs, onions to pull young. Leeks, turnips, spinage, caraway, basil and marjorum *on heat.* Plant potatoes, slips of tyme, lavender, sage, rosemary, rue, tansey, balm, hyssop, tarragon, wormwood, sorrel, savory, by parting of the roots; *mint,* by cutting the young sprigs about an inch in the ground, with a portion of the roots. Plant lettuces, celery for early use, cauliflowers, cabbages if required, leeks, turnips, transplant onions, prick out the capsicums sown last month on gentle heat; prick out celery sown in February or March. Hoeing and weeding are now required among all the crops planted in autumn, and as spring-planted grow, weeding and hoeing must be done to promote growth. Stick peas.
FRUIT GARDEN – Finish planting trees and shrubs and stocks;

head down newly-planted trees and stocks; finish hedging, ditching, and banking, all clean up. Sow fruit-stones, kernels, and pips, protect blossoms, rub off all the useless and foreright shoots of wall-trees in time. Attend to newly-grafted trees, and keep the clay firm. Sow seeds of forest trees.

FLOWER-GARDEN AND SHRUBBERY – *Sow* the same as last month, sow at the end for succession, the same things; but those then sowed in a hot-bed, to forward them, may now be sowed in the open ground: propagate, by layers, slips, and cuttings; strike slips or cuttings of China roses on gentle heat; prick out on gentle heat tender annuals to forward them. Shelter choice auriculas in pots from wind, rain, and sun. This is the *best time to plant evergreens;* plant dahlias, chrysanthemums, stocks and other hardy annuals. Mow grass-plats, etc, sweep and roll.

FORCING GROUND – Keep the grapery to about seventy-five degrees, pull off all useless shoots and top-bearing branches; keep your cucumbers and melons in free growth, less artificial heat, air and water more freely. If the plants droop at hot sun, shade for an hour or two, make beds for succession as required; *sow* cucumbers for planting under hand glasses in May; gourds, squashes, pumpkins. Keep the hardier things that are forced, such as potatoes, French beans, etc, duly watered; give air by taking the glasses off, cover at night.

GREEN-HOUSE – Fire heat will not now be wanted, unless to expel damp and sharp frost; air and water more freely. Shift and re-pot those plants which need it. Propagate by the rules laid down for last month. If insects appear, fumigate them, strike heaths, *sow and graft camellias;* prune and tie where wanted.

MAY

KITCHEN GARDEN – Sow kidney-beans, brocoli for spring use, cape for autumn, cauliflowers for December; Indian corn, cress, cucumbers under hand-glasses, and in the open ground, for pickling. Onions to plant out next year as bulbs; radishes, spinage, salsafy, skirrets, squash, nasturtiums, herbs, endive (not much), turnips, cabbages, savoys, lettuces, coleworts, prick out and plant celery, lettuces, capsicums, basil, marjorum and other annual herbs, love-apples; slip sage, this is proverbially the best time of the year for its striking. Plant radishes for seed, spring-sown cabbages, finish planting of potatoes; stick peas, move cucumbers and squashes put out last month, top beans when in blossom; hoe and thin out the crops of onions, carrots, lettuces, parsnips, and other spring crops; hoe and earth up peas, beans, potatoes, keep the hoe well moving, and destroy weeds *everywhere*. Tie up lettuces and cabbages to heart in, and blanch for use. *Mark the stumps* of best cabbages when cut, to put out for seed; when the stumps have sprouted above two inches, take them up, and lay them in by the heels at a foot apart, and about three feet from row to row.

FRUIT GARDEN – Look to your grafts, and take off or loosen the bandages as wanted, disbud or take off with your finger and thumb the foreright and superfluous young shoots of wall-trees; thin the fruit set on the apricots, peaches, and nectarines; nail in the vines as soon as possible. *Caterpillars will now be hatching.* Water new-planted trees if necessary, keep young stocks and seedlings clear in the nursery; if troubled with insects, water or rather pelt wall-trees with strong lime or tobacco water, or fumigate them with a bellows. Look over newly-budded and grafted trees, and rub off all the shoots that arise from the stock, below the bud of the graft. *Water straw-berries.*

FLOWER GARDEN – Weed, hoe, and set out to their proper distances, seedlings just come up; at about the middle or end of the month sow again for succession, larkspurs, mignonette, ten-week and wallflower-leaved stock, minor convolvulus, Virginia stock, propagate by slips, cuttings, and layers, double wallflowers, pinks, sweet-williams, rockets, scarlet lychnis, etc; thin seedlings soon to prevent their being drawn up spindling; pot out tender annuals, freely give air to plants in the frames as the weather permits. Shift and tie potted plants that need it;

keep all clean, remove auriculas in pots to a north-east aspect; take up bulbous and tuberous roots as their leaves decay. Keep the gravel-walks and grass in neat trim.

FORCING GROUND – Continue the grapery at the same heat; as the fruit swells, thin the berries of the bunches with scissors; water the vines with drainings of a dunghill. Less heat and covering will now be required for cucumbers and melons, and more air and water required; ridge out cucumbers under hand-glasses; water mushroom beds if dry.

GREEN-HOUSE – Air and water abundantly; towards the middle of the month remove the more hardy plants out into their summer stations, let the rest follow as occasion offers; very tender ones may still remain. Shift and propagate as before, bring in tender annuals from the frames.

JUNE

KITCHEN GARDEN – Sow kidney-beans, pumpkins, tomattos, coleworts for a supply of young winter greens; under the name of plants, they are sold nearly all the year round, in the London vegetable markets. Cucumbers for pickling, this is indeed the safest and perhaps the best time for sowing this article. It is so tender that it can but seldom stand the natural air in this climate till this month, many of the London market gardeners sow from two to twenty acres; their method is to sow it in shallow drills about six or eight feet apart, dropping the seeds about three or four inches apart in the drills, which are slightly watered if the ground is dry, if the weather is hot, the seeds vegetate and come up in a few days; after they get a rough leaf, they are thinned by hoeing them out to a foot or fifteen inches apart, and earthed with a hoe as they grow, and soon cover the ground; excellent crops are sometimes had from those sown even late in June, while those of May have been stunted and cankered; they seldom succeed by transplanting to the open air. Sow black Spanish radishes for autumn and winter use, other radishes if wanted, endive, principal sowing late in the month. Lettuces, the hardy-cosses are now the best to sow, celery for late, turnips, peas, cardoons. Plant cucumbers and gourds, pumpkins, nasturtiums, and in general similar articles not planted out last month, leeks, celery, cauliflowers, brocoli, borecole, and greencole, savoys, and other articles of autumn and winter use, seedling and struck slips of herbs, water them when wanted, hoe, thin out, and clean all from weeds of the spring-sown crops; earth beans and peas, potatoes, kidney-beans, top beans as they blossom, slipping of herbs will still succeed, tie up lettuces to blanch for use, stick peas, leave off cutting asparagus about the twenty-fourth, keep the hoe well employed, save some of the best and earliest cauliflowers for seed, cut mint and other herbs for drying; a general rule for cutting herbs for drying, is, to cut them when in full flower.

FRUIT GARDEN – We have little labour here now, except in the prospect of gathering the crop as it comes in; tie up and secure young grafted trees, trimming the stocks of the wild wood; summer prune and nail wall trees. Budding may be begun at the close of the month. Clip hedges, net cherries.

FLOWER GARDEN – General work much the same as last month; tie tall-growing flowers up to sticks; sow Brompton, Twicken-ham, and giant stocks to flower next spring, lay roses, evergreens, slip myrtles to strike, pipe and lay pinks and carnations, plant tender annuals in borders, plant out in nursery-beds seedling pinks, carnations and pickatees* (sown on slight heat in April), auriculas and polyanthuses in shady places.

FORCING GROUND – Finish thinning grapes in grapery, keep the vines neatly trimmed and tied, discontinue fires, unless in cold or damp weather, still attend to cucumbers and melons, give air and water freely, attend to the succession crops, let the vines of the cucumbers planted under hand-glasses, run out from beneath by tilting the glasses, prune them occasionally.

GREEN-HOUSE – Give air and water as wanted. Those plants that are out of the house must be attended to, watered when necessary, trimmed and tied; if excessive rain falls for days

*Picotee. = a variety of carnation (O.E.D.)

together, and you fear too much moisture, turn the pots sideways, set out what were too tender to put out last month. Propagate as last month.

JULY

KITCHEN GARDEN – Sow kidney-beans for the last crop about the twentieth, they seldom succeed if sown later, early dwarf cabbages, a principal sowing, to plant out in October, for the general crop for next spring and summer's use may be made from the twenty-fifth to the thirty-first, endive for autumn, peas and beans have still a little chance of success; radishes, lettuces, only the more hardy sorts will now succeed; onions a few to pull green in the autumn for salads, it should be Lisbon or Reading onion. Coleworts for a main crop for winter, early in the month, turnips principal sowing of the year, for autumn and winter use, chervil to stand the winter. Plant celery, endive, lettuces, cabbages, leeks, savoys, brocoli, greencale, cauliflowers. Hoe and keep all clean; dry herbs, pull up, dry, and house, onions, garlick, shalots and the like as the tops fade. Stick peas and scarlet-beans, blanch white beets, tie up lettuces, and endive for blanching, top beans, earth celery, gather seeds.

FRUIT GARDEN – Bud; water, if dry, newly-planted fruit trees; nail and thin, and trim wall-fruit-trees, keep all in neat order. Head down young espaliers; stop fruit-bearing shoots of vines; prune away shoots and suckers from the stems of trees. Net morella and other cherries, and currants.

FLOWER GARDEN – General work as last month. What was not done then, must be done this; repetition would be useless; part auriculas and polyanthus roots, gather seeds, dry and house them, mark every sort and sample. Plant saffron crocus and other autumn bulbs. Sow mignonette in pots to blow in shelter in the winter, likewise ten-week stock, both at the end of the month, to make an earlier sowing as well, would give a better chance. Clip box, also evergreen hedges.

FORCING GROUND – The crop in the grapery will now be ripening or ripe, keep the vines neatly trimmed; give plenty of air to colour the grapes; a fire may be necessary in very damp weather to prevent injury to the fruit. Successional graperies require the same treatment in the different stages of their growth. Attend to your late cucumbers and melons, of which, if properly managed, you will have abundance.

GREEN-HOUSE – General treatment as last month. Geranium cuttings will now strike like weeds in the open ground, from which they are easily potted. Strike heaths. This is the best month, particularly the early part of it, to strike myrtles. The slips should be about two inches long, thickly pricked out into large pots, or under a hand-glass on gentle heat, and shaded from hot sun, till they have struck, which will be seen by their making growth; they will not be fit to pot till September.

AUGUST

KITCHEN GARDEN – Sow early cabbages in the first week, the last sowing for the year. Red cabbages same time; cauliflower for spring and summer use, about the twenty-first, cress, hardy coss and cabbage lettuces from about the twelfth to the thirtieth, to stand the winter, prickly spinage for a principal winter crop about the same time. Turnips last sowing, Welsh onions first week in the month, Lisbon, Strasburg, and Reading, from the middle to the end of the month. Radishes, chervil, Spanish turnip radishes, carrots to pull early in the spring, corn salad, plant endive in sheltered spots; planting for winter must not now be delayed, put out as fast as possible, if not enough planted, savoys, cabbages, brocoli, coleworts, celery etc. Hoe, and earth up where necessary, weed, and thin young crops, water and shade where requisite. Gather seeds as they ripen, dry and store them, dry and house onions, garlic, shalots; dry herbs as they flower. Dig up, dry, and put by mushroom spawn.

FRUIT GARDEN – Finish budding; and loosen the bandages round buds, put in earlier, or take them entirely off, plant out strawberries that struck in the fore part of the spring; keep all the wall-trees in neat trim; early in the month wattle round with willow or hurdle rods some of the best currant-bushes, where none of the fruit has been fingered, and mat them over, or, what is better, cover them with canvass or bunting; the crop may with care be preserved till October.

FLOWER GARDEN – Continue to take up, dry and store, bulbous and other flower roots, as the leaves decay. General work as last month. Gather seeds of shrubs and flowers. Sow in pots for succession to blow in the winter. Plant autumnal bulbs; plant where to remain, pipings and layers, and seedlings of hardy plants; transplant auriculas and their like. Shift succulent plants in pots.

FORCING GROUND – The greatest part of the labour here is now over for the year. Attend to what you have left, water and thin as required; make mushroom beds, put by and repair lights and frames as they fall into disuse; or place them to forward and protect wall-fruit.

GREEN HOUSE – The same management as last month and the previous, pot and shift towards the end of the month. Pot all your young stock, raised from seed, cuttings or layers.

SEPTEMBER

KITCHEN GARDEN – Sow Reading onions for transplanting in the spring, carrots on warm border; both the first week; spinage, Spanish and other radishes in warm spots, same time. Plant, for spring, coleworts, savoys for greens, late brocoli, celery, lettuces, and endive on warm borders, or, dry open pieces; herbs, culinary and medicinal; prick out cauliflowers; clear herb-beds and all decayed articles; finish drying and housing of onions, seeds, etc, earth up celery, cardoons, take up and house potatoes, cut onion seed.

FRUIT GARDEN – Propagate by layers and cuttings, gather and store keeping fruit as it ripens or is fit; net grapes, bag the best in bags of gauze, crape or bunting. Keep all trim on the wall-trees; thin the leaves (but not too much), where they impede the rays of the sun from the fruit; prepare ground for planting fruit-trees.

FLOWER GARDEN – Take up tiger flowers and other tender bulbs. Put in pots, Guernsey and Belladonna lilies: clear decayed flowers and littered leaves away; trim plants and shrubs, the best auriculas in pots, dress, shift, and place in shady shelter. Towards the end of the month plant crocuses, common anemones, early tulips, lilies, and scaly bulbs: remove succulent potted plants into shelter, as aloes, Indian figs, etc. Remove potted mignonette to a warm sunny border, or place it in frames, covering at night. Plant out seedling carnations, pinks, and pickatees, also layers and pipings of, where they are to remain; gather seeds.

FORCING GROUND – The observations of last month are applicable to this, almost all is in a state of decline and inactivity here, as far as regards forcing; the same may be said of the succeeding month and November, except that in November preparations may be made for forcing asparagus and seacale; this month, however it is a good time for making mushroom beds.

GREEN-HOUSE – Ha! we smell winter here. General management the same as last month. The more tender plants should now be taken in the first week; it is better to be too early than too late; and if surprised by frost, the effects are woeful. The whole should be housed by the end of the month as a measure of prudence.

OCTOBER

KITCHEN GARDEN – Sow a few mazagan beans, cress, white coss lettuces in frames for spring planting. Manure well, and plant the principal crop of cabbages for spring, lettuces, coleworts, celery, last planting; shalots, garlic on dry ground, strawberries, bulbs, herbs, clear off decayed leaves, hoe and weed, get all as clean up as possible; finish taking up and housing potatoes, the same of carrots, salsafy, scorzenera, shelter seedling cauliflowers from frost. or they will be black-shanked. Break in the leaves of the late cauliflower that show fruit, to cover them from frost; early white brocoli, the same, cut down asparagus-stalks, trench vacant ground, manure. Get all your planting close up.

FRUIT GARDEN – Transplant young fruit-trees; plant stocks of all kinds as soon as the sap is down. Apples and pears, and all winter-keeping fruit, must not be delayed gathering beyond this month. Sow cherry-stones, take up layers of trees and shrubs of last year's laying and plant them.

FLOWER GARDEN – Very hardy flowers, such as larkspurs, may be sowed to stand the winter, and blossom early, plant hyacinths, tulips, anemones, ranunculuses, bulbous irises, etc. Finish separating and planting, carnation layers, etc. Put hyacinths, narcissuses, early tulips and jonquils, in water-glasses. Plant cuttings of jasmine, laurel, cubas, honeysuckle, etc, preserve potted carnations from wet and frost, place them in a sunny situation, house or put under glass the mignonette in pots. Plant deciduous trees and shrubs, herbaceous plants.

FORCING GROUND – The directions for last month apply here, attend to your mushroom beds.

GREEN-HOUSE – Water sparingly, give much air, move all in, arrange your plants and take care of them.

NOVEMBER

KITCHEN GARDEN – Sow early peas, leeks, beans, radishes, and early horn carrots, on warm borders, salad on heat. Plant cabbage-stumps for seed, cauliflowers under hand-glasses and frames early in the month, lettuces under cover for winter use, endive, rhubarb in rows after parting old roots; finish taking up potatoes early, or the frost may save you the trouble; take up and store carrots, beets, some parsnips also, to get at in frost. Thin lettuces sown last month in frames, sift fine dry mould among them to strengthen them, keep them dry, let them not have the least shower, all depends on keeping them dry, yet keep the lights off during the day, and tilt at night to give as much air as possible unless hard frost. Give free air to cauliflower in frames and glasses. Examine onion and other stores.

FRUIT GARDEN – Sow the fruit stones that have till now been kept in sand; cut old wood from raspberries, begin pruning. Plant wall and other trees, dig among trees, plant stocks.

FLOWER GARDEN – Pull up dead flowers, and tie up those that blossom where wanted; protect tender flowers, auriculas, etc, from wet and frost. Plant bulbs, hardy biennials and perennials, shrubs and trees, as last month if dry weather; prepare shrubs for forcing, as China roses, Persian lilac, etc. Attend to the grass lawns and verges, sweep up leaves, and pole down worm casts; turn gravel-walks if weedy or mossy. Take up dahlia roots, and place them secure from frost.

FORCING GROUND – Attention required as last month, but nothing new to be done, except making beds for asparagus; and forcing seacale.

GREEN-HOUSE – Air and water as required, but more sparingly; prune and clean, pot bulbs for forcing.

DECEMBER

KITCHEN GARDEN – Sow radishes on heat, or on warm borders, cover peas and beans. Planting ought to have all been done before. Weed, hoeing is now of little use; protect from frost endive, celery, and what you can. Attend to lettuces and cauliflowers under glass as last month; tie endive when dry; earth up all celery pretty closely. Hedge, ditch, and drain as wanted, dig, trench, and manure vacant land.

FRUIT GARDEN – Whatever was done last month, may be done this, except pruning, which may as well be deferred, according to convenience.

FLOWER GARDEN – Nothing can be usefully sown or planted; protect tender shrubs and plants by matting them, or straw, cover bulbs from frost, dress and dig flower borders, and shrubberies; protect flowers as last month.

FORCING GROUND – All is nearly dormant here, force asparagus and seacale, prepare for sowing cucumbers.

GREEN-HOUSE – Protect by fires from damp and frost, give air on mild days.

FARMING

FARMING

COBBETT'S FARM

18 December 1830

The public knows what the Bloody Old *Times* and the *Bull-frog's Blunderer,* commonly called the *Farmer's Journal,* have put forth about my farm, they have, and particularly a malignant and stupid wretch under the signature of T., said that my land was all a *bed of weeds,* that my *swedish turnips* (that were on the land last fall) were small and like so many plants of grass, so thick on the ground; that I had no straw on the premises; and that *everything had the hue of misery itself.* This is a species of calumny that even this infamous press never resorted to in the case of *any other man.* As to the base people who conduct these newspapers, though they deserve to be knocked on the head and left on the highways or on commons for the carrion-crows to eat, they are not worthy of my vengeance; it is their infamous *setters-on* and *backers-on* that call for my vengeance; and on these I inflict it whenever I can, and rejoice when I see it *inflicted by others.* The Bloody Old *Times* attacks me on behalf of the loan-mongers and stock-jobbers and Jews, and the *Bull-frog's Blunderer* attacks me on behalf of the bullet-headed, monopolising, greedy, grinding, cruel Bull-frog farmers and their equally cruel landlords and parsons. Therefore, *on these* my vengeance is due, and on them I inflict it, when I can, and I always rejoice when it is inflicted on them by others. It is not the *stick,* but the *hand that wields* it, that the sensible dog always bites. To put on record a clear proof of the malignity of these base people, the best way will be, to give an *account of my crops at the end of my lease*. If I were to say anything about the *state of the land*, there might be room for difference of opinion; but the *amount of the crop*, and that amount to be verified by reference to a third party, can leave no doubt. The following, then, is the statement, in answer to the bloody tool of the Jews and the base tool of the *Bull-frogs*, the whole of whom *I* challenge to equal the crops of this calumniated farm:—

The meadows were grazed during the year.
The arable land consisted of
Wheat – 5 acres, rather short measure.
Barley – 18 acres.
Potatoes 6, which Mrs. Cobbett (I being away) was persuaded by some of our gardening neighbours, to believe a source of immense wealth. However, as she was *farmer*, she had a right to do what she pleased with the land.
Swedish Turnip seed   13
Swedish Turnips   6
White Turnips   4
Mangel Wurzel   8

And this was the *whole of the arable land*. The produce was as follows:—

Wheat 25 quarters, besides a sack to an acre, which all the care in the world could not prevent the *birds* from eating.
Swedish Turnip seed, 320 bushels.
Barley 108 quarters.

Only a part has been thrashed out yet. The wheat has been sold for 78*s*. a quarter, the barley (wet a little part of it) for 38*s*. what has been sold. The potatoes were sold for 12*l*. an acre, being compelled to get them off the land by Michaelmas-day. I estimate the barley, it not being all thrashed yet; but, I am sure that I am under the mark. The wheat was the handsomest piece of wheat I ever saw in my whole life. It was what we call, in Hampshire, *the old-fashioned white-straw:* it has white straw, white ear, and *brown* small grain. It is the very sort of wheat that is grown in Virginia, and that makes the finest flour. A neighbour, who sold me the seed, got it from Wiltshire. Every body said that they never saw such a piece of wheat. The eight acres of barley that grew in the same field, Sir·Thomas Beevor saw in full ear, and said it was *the finest piece of barley he had ever seen in his life,* though all his life, observe, he has been in Norfolk. The other ten acres of barley grew where there had been *two crops of Cobbett's Corn succeeding each other.* The Swedish Turnip seed had succeeded as fine a crop of mangel wurzel as ever was seen. The potatoes had succeeded a crop of cabbages much finer than ever I saw before. The potatoes had 12 loads of manure to the acre; but all the other crops not only had no manure *for them,* but *the land never had any while it was in my possession.* The manure I found on the farm was not sufficient for an acre of land, and I left, in the yard, enough for the 13 acres where the turnip seed had stood, and that land I left ploughed and harrowed, and ready for wheat. The turnips and mangel wurzels left for winter consumption on the land, had been put on manure to the amount of 12 loads to the acre. So that there were 24 (with the potatoe-ground) well manured, and manure for 13 acres more; and never was farm *left* in such beautiful condition. But *the crops* speak for the farm. The scoundrel T., one of whose paragraphs my foolish landlord had the emptiness to send to me, said that my Swedish turnips were *thick on* ground! The base, broken-down sponger, whom I know very well, and whose name I keep out of print for the sake of those who have the misfortune to be related to him, saw those turnips on ridges *six feet apart,* two rows on a ridge a *foot apart,* and the plants a foot apart in the row; and this is what the stupid and malignant villain called *thick upon the ground!* This piece of seed was the finest, I believe, that ever was seen. The land, between the plants, was *ploughed in the winter;* then

twice *in the spring*. The seed-stalks were *six feet high;* the flower
of the Swedish Turnip is a very beautiful pale yellow; and the
piece being so large, it attracted all eyes that came near the
spot; people used to stand in groups on the Thames-bank and
look at it. But, did ever any farmer before grow 320 bushels of
Swedish turnip on 13 acres of land, in one piece? I never heard
of such a thing; and the finest of seed too; and the land a fallow
at the same time. The *wide distances* let in the air and sun; and
the inter-tillage *fed the plants* so as to *fill the top-pods of the branches.*
A fact worth the attention of farmers is, that the seed from which
these plants came was at least *seven years old!* It was a bag of
seed given to me, in 1822, by Mr. Peppercorn, who then rented
a farm of the Whitbreads, *in Bedfordshire.* He gave it to me as
being extraordinarily fine. I had tried it in my garden, and
found it to be so; but never had an opportunity of raising a lot
of seed from it before. There was but *one single false plant* in
the whole piece, and that I had pulled up before the bloom
appeared. The seed is the finest sample that I ever saw. I
shall, by and by, offer this seed for sale, in any quantities,
great or small; and also *mangel wurzel seed,* grown by me, the
autumn before, from plants of the *deep red* sort, carefully
selected for the purpose.—Besides the seeds, I have some *locust
seed;* and these are all the seeds I have, or shall have, for sale.
Of apple and pear trees, and some others, I have a good parcel;
but it is too troublesome to be selling these in the winter; I shall
therefore, not attempt it till towards March.

I make no apology for this article. It contains useful information. In the first place, it is useful to show up the baseness
of the tools of the people's worst enemies; and, in the next place,
here is information that may be useful to many persons
who cultivate the land, and particularly to such as raise
seeds in great quantities. It is, too, of the greatest importance, that it be known that Cobbett's Corn will succeed
in *any summer* in England. I have the crop to *show* to any
one that may choose to see it; and this is the best of answers
to all gainsayers.

WANTED

I WANT a young man from 16 to 18 years of age, as a sort of *farming* and *gardening* and *nursery-work* APPRENTICE. The case is this: my farm is taken care of by my only surviving brother, who has been either gardener or farmer all his life time, and who, though he is only thirteen months older than I am, is not able to move about so quickly, from place to place, as is necessary in a concern like mine; nor is it proper that he should be exposed to wet and cold so much as is absolutely necessary to the due looking after of ten or a dozen men. He has, from the fruit of his own labour, raised a family of ten able and good children, who have already brought him about a score of grand-children. I know that he began to work *hard* more than *fifty* years ago; and verily I believe him to have done more hard work than any man now living in England. While his limbs feel the consequences of this, his experience and skill remain; and I want a young man, or lad, of the age above mentioned, to supply *legs*, which my brother cannot move so nimbly as the case requires. He is not wanted to *work much;* but, if he learn to do every thing on a farm, it will do him no harm. He is wanted to see the orders of myself, or of my brother, obeyed. I do not want a *Bailiff;* for bailiffs do not think that they earn their wages, unless they furnish you with *science,* or, at least, with *advice,* as well as with care; and I want neither science nor advice. I want *legs* that will move nimbly and willingly, and a young head capable of *learning.* The lad ought to be *stout,* and not stunted; he ought to be able to read and write a little; but, two things are indispensable; namely, that his father be a *farmer,* and that the son has lived on a farm in *England,* all his life, and at a distance of not less than forty miles from London; and not less than twenty miles from Portsmouth, Plymouth, Bath, Bristol, Cheltenham, Liverpool, Manchester, Leeds, or Norwich. He is to sit at table with my brother and my niece, (who is the housekeeper), and, when I am at the farm, with me also; and is to be treated in every respect as *the young farmer of the house.* He is never to quit the farm; except on my business, and to go to the parish church on the Sunday, and is to be under the control of my brother as completely as if he were his son. He will here learn all about cultivation of Indian Corn, Mangel Wurzle, and of several things not very common. He will learn to sow and rear trees, and to plant and prune them. He will learn how to raise seeds of various sorts. He will learn how to grind and dress wheat and Indian Corn, and will see how the flour is applied. He will learn how to make beer; to see butter and cheese made before breakfast time; and he will have constantly before his eyes examples of early rising, activity, punctual attention to business, *content with plain living,* and *perfect sobriety.*—Now, if any farmer, who is of *my political principles, full up to the mark,* have such a son, nephew, or grandson, to dispose of in this way, I shall be glad to hear from him on the subject. If the lad stay a year out, I will make him *a present* of not less than ten or twelve guineas. It is, I hope, unnecessary to add, that this is a farm-house without a *tea-kettle* or a *coffee-pot,* and without any of the *sweets* that come from the *sweats* of African slaves. Please to observe, that I do not want a *young gentleman;* but a good, sturdy lad, whose hands do not instinctively recoil from a frozen chain, or from the dirty heels of an ox or a horse. I hope that the lad, or young man, that I am to have, will never have been at an *establishment,* vulgarly called a *boarding-school:* if he unfortunately have, and should suit in all other respects, I must sweat the boarding-school nonsense out of him; that is all. If he have a mind to improve himself in study, here are books, and all the other means of well employing his leisure hours.—Letters to me on the subject to be *postage paid.*

Wm. COBBETT.

183 Fleet Street, London.

P.S. The *great qualities* are, a fitness to *give orders,* and *spirit to enforce obedience;* and, above all things, never to *connive* at the misconduct of the men; but invariably to make a true report of their behaviour, whether good or bad. It will be quite useless to engage a soft, milky thing, that has not the courage to make a lazy fellow stir, or to reprove a perverse one. Fathers will know what stuff their sons are made of, and will, of course, not recommend them to me, unless fit for my purpose. *They* know well *what I want,* and I beg them not to offer me what will not suit me.—The lad will, of course, be boarded, lodged, and have his clothes washed in the house.

N.B. I will have no one, who has any near relation that is a *tax-eater* of any sort.

Normandy Farm, 13 *Oct.* 1833.
ITALIAN CLOVER

THE following matter I think to be of very great importance to all persons having fields, gardens, or wishing to have plantations. Peers have all got fields and gardens; or, at least, I hope so; for I am sure that they ought to have some. The law compels us members of the House of Commons to be *landowners*; and, therefore, here I am sure of my mark. For these reasons I take the liberty, very respectfully, to request their Lordships, and also the members of the other honourable House, to read the whole of this article, when they can deduct from their other more grave and important occupations the time necessary for the purpose.

Mr Thomas Hayley, of *Liverpool,* has requested me to state that he has a quantity of the seed of this clover, which is called, by botanists, "*Trifolium Incarnatum,*" for what reason the "*Incarnatum*" makes part of its name, I know not, but I am inclined to think that it is a plant of great importance. Before I proceed to communicate what I know relative to it, I will insert an article on the subject, which I take from the *Plymouth Journal.*

Trifolium Incarnatum.—This is a species of clover from Italy, much approved in this country, and which promises to be of great advantage to agriculturists; it bears a beautiful head of bright red flowers, resembling sainfoin in colour, and requires a good soil. The mode of culture hitherto pursued has been to plough up a wheat stubble immediately after harvest, sow the seed, say 10lb. per acre. It produces a large burden, which comes to use at the commencement of the following May. A period when a supply of green food must be of incalculable value, and which will admit of a crop of turnips following in succession. It has lately been sought for with the greatest avidity, and is likely to get into general use.

Now, being at Mr. Smith's, at *Languard* farm, in the middle of July, 1832; and the pleasure I enjoyed there will not be presently forgotten; for there I saw a real English farm house, and two pork tubs, containing each, I should suppose, three quarters of a ton of pork, to be eaten by those who made the wheat and the pork to come: being at this farm, I saw, in the garden, a patch of this *Italian* clover. I did not inquire at what time of the year it had been sowed; but I will obtain that information next week. It must have been fit to cut up for horses in May; for it was so ripe in the middle of July, that the seed might have been thrashed out at that time. Nay, some of it was; for I cut a small parcel of it, and beat out the seed which Mr. Smith was so good as to give me. Far advanced as the plants were, and hard and sticky, my horse ate them with great greediness; and I, therefore, took some of the seed, in order to try it the next year. It was sowed in my garden at *Kensington*, or, at least, part of it, late in August of that year; but as I had to give up that garden last Christmas, the plants were, of course, destroyed; but, they grew very vigorously, promised a most abundant crop; and I am very sure that they would have produced excellent food for horses in May. There is this to be observed, however, that the ground was as rich as it could possibly be; and that the seed was sowed with the greatest possible care.

Mr. Hayley having written to me as above, but not having

sent me a sample of the seed, I wrote to him for such a sample. By the time that the sample came, we were full six weeks too late for the sowing, according to this *Plymouth* account; but I took a little of the seed, put it under a frank or two (I wish every frank of mine were as usefully employed!), sent it down into Surrey that a little patch might be sowed in that early ground; and it was sowed accordingly about six or eight days ago. It is very curious that this led to the discovery that I already knew the plant; for, my son having in his possession the remainder of the seed which I got from Mr. Smith, compared it with this seed from Mr. Hayley, and found it to be the seed of the same plant; so that, without my being at all aware of it, I had already great knowledge of this *Trifolium Incarnatum,* which I must call *Italian clover;* for, if I were to persevere in the use of this botanical name. I defy any human being to conjecture what it will be called at last, by those who will have to sow it and use it.

This plant, unlike the common clover, the lucerne, and the sainfoin, has a bushy, fibrous, and spreading root; and is very likely to be an *annual;* and it is certain that turnips, or mangelwurzel, or the Cobbett-corn, would very conveniently follow it on the same ground, during the same summer. I think it very likely that it may require rich land; or, at least, land well tilled, clean, and sufficiently rich for manure; but, what land can be too costly that will produce a large crop of green food for horses and cattle in the month of May; when the horses especially want green food; when the clover is not come; and when even the lucerne (if you have the peculiar ground which it requires) is only beginning; and when tares and winter barley are so *flashy,* so watery and unsubstantial, as to do horses harm rather than good? Plenty of all these things in the month of June; but none of them, except the lucerne, in the all-pinching month of May; which, though chanted by poets, is not, I can assure their worships, in whatever garret they may be seated, chanted by the horses and the cows and the sheep; for then the turnips, the carrots, the mangel-wurzels, are all gone; as to the grass, though it begins to sprout and look gay, to let their mouths touch it, destroys the hay-crop. Therefore, the beautiful herbage is forbidden fruit to them. The *birds,* indeed, do chant, and the hawthorns blow; but even here their worships, the poets, should take a little care of what they are about; for, as the old country people say, "The Parliament changed the *style,* but they could not change the *seasons.*" And I can remember when the old people spoke very *resentfully* of this attempt at what they thought to change the seasons. The bloom of the hawthorn is emphatically called *May;* but except in very warm situations, and in these southern counties too, it never does come out into bloom till the first of May, *old style.*

The only question to be settled with regard to this *trifolium incarnatum,* or *Italian clover,* is this, *whether it will stand the winter in England.* It is to be observed, that the trials, of which the *Plymouth Journal* gives an account, must have been made, in all probability, *in Devonshire;* and that the trial made by Mr. Smith was in the *Isle of Wight.* Mr. Smith's seed must have been sowed in the fall, and must, therefore, have stood the winter, which was not a very mild one. It is possible that this plant would not stand the winter in some parts of England; but if it will stand the winter in the southern part of it only; cutting the country across at *Meriden;* or, if it be confined to the counties south of the *Thames.* This cannot be; but, if it were, it is a matter of the very greatest importance. My horse had plenty of corn and plenty of clover, at the same time, yet he ate up this *Italian clover* with great eagerness.

I have not yet written to Mr. Hayley about the best mode of disposing of the seed. Mr. Hayley appears to be a merchant, and to have received this seed as a consignment from his correspondent in Italy. I shall advise him, by next Saturday's post, not to sell the seed, except at such a rate as shall make it necessary for seedsmen to sell it by *retail* at half-a-crown a pound, he making, as an allowance to them, the usual allowance given to seedsmen, or the usual profit which they are in the

habit of having, which is very large, and which every reasonable man will see ought to be very large, their commodities
not being like candles, soap, and sugar, and yards of cotton
and the like, which need nothing but a shop-window, a
populous neighbourhood, and a person standing behind the
counter capable of counting twelvepence, and of weighing
sixteen ounces.

I shall, for my own use, beg Mr. Hayley to let me have upon
some terms or other, a parcel of this seed; and if he have a
mind to have some of it sold at *Bolt-Court,* which is very likely,
I will apprize my readers of it, when I receive the seed, which
I strongly recommend to gentlemen to try upon a small scale,
under their own eye. It is too late to make the trial this fall;
because it will now take some time to get the seed to London;
but, I am so strongly persuaded of the great value of the plant,
that I shall take care to provide myself with some of the seed;
and other persons may do the same, if they choose. In the
spring I shall be able to speak positively about its capability
of standing the winter.

Now, seeing that the editor of the *Morning Chronicle* has
recently anticipated, with so much joy, the *close of my life,*
natural, political, and literary, and has accused me of getting
a sort of shifty maintenance by *much bad bookmaking,* let me just
ask, whether he can, in the whole course of his life, possibly
have rendered so much service to England, as it is very likely
that I shall render by the writing of this one single article.
To be sure, it is not I who have introduced this plant into
England; for that we are indebted to somebody else. But, who
besides me, would, upon seeing the plant grow in the *Isle of
Wight,* have taken care to try his horse with it; have brought
away some of the seed; have had that seed sowed in a few days
afterwards; have preserved the rest of the seed, in order to be
tried afterwards; and, above all, who was there to cause a
knowledge of the plant thus to be communicated to the whole
nation at once, and that, too, in a manner which must force
the subject upon the attention of very great numbers of the

most sensible men in the kingdom and such, too, as have the means of verifying the facts relating to this very important matter? let me ask this; and then ask the reader, whether there be anything too severe, in the way of literary punishment, which I can inflict upon this envious and spiteful wretch? Thus I leave this subject for the present. When I get an answer from Mr. Hayley, I shall give my readers intelligence about the disposal of the seed, except that I can tell them now, that there will be none of it to be sold at *Bolt-court* under half-a-crown a pound.

ITALIAN CLOVER SEED

In consequence of my publications on the subject. I have received a letter from Lord Vernon, in which is the following sentence: 'I have been cultivating the *Italian clover* for two or three years, and have every reason to believe that it will stand the frost."

So that this important point seems settled. I shall sow a piece in April, and have no doubt of a crop to cut up for horses in June; but, if sowed in August, after wheat, or any other grain, it will doubtless be fit to cut up for horses in May, or, in the south of England, the latter end of April. Sowed after the last ploughing between ridges of Tullian wheat or barley, it would be a crop, yielding a great quantity of food before November. There is every appearance of its soon finding its way all over the country, in the causing of which I shall be very proud of having been an humble instrument. A gentleman at *Newcastle* tells me, that the Italians apply the epithet *Incarnatum* to it, on account of its large ruddy and *flesh*-like flower. But we must call it 'Italian clover,' or we shall never know what we are talking about. The chopsticks will very soon call it *Talin* clover; and if we were to take that name at the beginning it would be just as well. It will grow just as well, and bear just as good crops, without the assistance of *heddekashun*.

COBBETT'S SOW

I have to announce that I mean to do, what I never did before; that is, to send a fat hog to the *cattle* show at Smithfield on Monday next. She (for it is a 'female') has by no means finished her fatting; but I am of opinion, that she is now, the finest specimen of a fat hog, that ever was seen in England or anywhere else. She far surpasses every thing, that ever I saw; and I challenge all the *Bullfrogs* in England to produce any-thing to equal her. I mentioned before, that this sow came up in a crate from Worcestershire last April. She was so small as to be put in at a wooden door made at the top of the crate; and when she is fat, I will bet a hundred pounds she will weigh *twenty five score* or *five hundred pounds*, a weight much about equal to that of a fat Scotch ox.

Whether I shall find too much difficulty in getting her into a cart, I know not. If I do, she can not go; if I do not, go she shall; for she is really a sight worth the beholding of any man. Valued very highly as a store pig, she was in September last worth *three pounds*. She has not yet cost *forty shillings* in the fatting; and, at the price at which Hampshire hogs sell at this moment *down in the country*, she is worth *twelve pounds ten shillings*.

If I should not be able to send her to the Smithfield show, I hereby invite all graziers, salesmen of cattle, carcass-butchers, and all bull-frogs, and all farmers, of course, to come and see

her, and mark her manners of going on. To my great misfortune and irreparable loss, she, to a certainty, has been
reading the works of Malthus and Peter Thimble, as published
by Carlile; and thus corrupted, she seems to have set her face
so heartily against all the propensities tending to the propagation of her species, that I have been compelled to deliver
her over or to make preparations for so delivering her to that
sort of punishment, which the corrupters of mind would receive,
if they had their due.

PLOUGHING
Oxen v. Horses

I must give my opinion relative to the choice between oxen and
horses, in this business of corn;* and, indeed, in farm-work
generally. I am decidedly for the using of oxen, for reasons a
bare list of which, each occupying one line, would fill a page of
this book. In the FIRST place, the harness, if harness be used and
not yokes, and I will proceed first upon that supposition, is
much less expensive, and requires less strength, than that which
is to stand the jerking and the starting of a horse. SECOND:
food upon which a horse will not be able to work at all is quite
sufficient for an ox; the latter does not cost for his food a fourth,
or even a sixth and perhaps a tenth, taking the whole year
round, part of the sum which the food of a horse costs; and, of
the sort of food for an ox I shall speak by-and-bye. One of the
great plagues of horses is the blacksmith, who may almost be
looked upon as an inmate of the farm-yard, acting, as he
generally does, in the double capacity of horse-shoe-maker and
farrier, in the former of which, he, several times in every year,
actually makes business for himself in the latter. In short, this
may be called an everlasting visitor; and, being a prowler
about from place to place, he brings all the news regularly,
once or twice a week; and gathers a goodly group about
him at the stable door. Then, just at the time when you want

*Sweetcorn, or maize

the team to go out, a horse has got a shoe loose; he must be taken to the blacksmith, at perhaps a mile distance; or the blacksmith must be brought; and he, unluckily, is gone to another farm. How often does it happen (and every farmer will say it) for a wagon or cart, which ought to be off before day-light, to be kept at home till eight o'clock waiting for the operations of the blacksmith! How often does it happen for a harvest-wagon, to stand still for hours from the same cause!

With oxen you have none of these plagues, and none of the heavy expenses that accompany them. THIRD: there is the farrier with his *balls*, and his *drinks*, and his *salve*, and his *tow*, and all his tinkerings about day after day, week after week, and month after month. There is the *grease*, and the *pole-evil*, and the *glanders*, and the *strangles*, and the *fret*, and the *coughs*, and the *staggers*, and the *botts*, and various other nasty and troublesome diseases. The ox knows none of these; he sets them all, Bott Smith's namesakes and the whole, completely at defiance. If he gets *lamed* by any means, you have only to let him lay in a rough field or a meadow and rest until he be well; and if the lameness be incurable, still he will fat with very little trouble, and will, nineteen times out of twenty, sell for more than he cost. The farrier's bill is a manuscript of considerable length, winding up with a decent allegation in figures. You will find not a single ball omitted; and, generally speaking, I say generally speaking, the cost of the farrier is far beyond the good that he does; and in innumerable cases, you have at last to send the horse to the dog kennel. FOURTH: a personage coming still more home to you; I mean the carter. A carter is the sole master of the horses with which he goes; and, in nine cases out of ten, he is, as far as concerns them and their labours, pretty nearly the master of their owner. He must have his way pretty much as to quantity and quality of food, as to hours of labour, and as to various other things, in which, if you do not give way to him, you must make up your mind to get rid of him; and, even then, you only exchange one sort of half master for another. If you be peremptory in your commands to him, and insist upon such

or such a space of time, and also insist upon having your own
way with regard to the food of the horses, he has a way of
making their rough coats and bare bones convince you, that
he understood these matters a great deal better than you.
With oxen you have no part of this everlasting plague. They
want neither currying, nor rubbing; they want no straw cut up
for chaff, they want no stables to be cleaned out, once or twice
a day; they want no careful racking up by candle-light; they
want no man in the stable, two hours before it is time to turn
out to work: turned into the field or the meadow, or turned to
the cribs in their yard, they are ready at daylight to receive
the collar or the yoke, and they are at work without any
previous ceremony. The carter gets drunk, or quits you, which
he legally may, in the middle of harvest, though he has been
living upon you all the winter; he may do this legally if he be
fired with the love of fame to be acquired in *his Majesty's
service*. With oxen you set both the carter, and this most
injurious law at defiance. Any thing that has two hands can
put oxen in harness, and can hold a plough, or drive a cart or a
wagon, with creatures so gentle. You have no growling fellow

to share in the mastership of the concern with yourself; and you expose none of these to act unjustly and ungratefully. FIFTH: food which the ox requires is, any tolerably good meadow or grass-field in summer, or, which is better, clover, lucerne, or even meadow grass cut up and brought in by himself, to be eaten out of a crib in the yard; because, especially in the busy time of the year, the ox gets his meat quietly, and lies down and rests. Here is no cutting up of chaff, no going to the granary for oats or beans; or, for what is a great deal better, corn: no bins are wanted, and there is no sharing between the working cattle and the rats and the mice, both of whom invariably, partake of the meal administered in the manger. In winter time, when there is no grass, coarse hay is sufficient for the ox, corn-tops, or even stalks, which latter may serve in the fall of the year, unless the oxen work hard. In times of leisure the ox is getting flesh, notwithstanding the cheapness of his food. White turnips, tops and all; mangel wurzel, and its tops; and, towards the spring of the year, when the work is required to be hard, and before the grass has arrived, or even the clover or the lucerne, *Swedish turnips* is the proper food for an ox. He would fatten on them if he did no work; but he will work on them constantly, without losing flesh. You know how many oxen you have; and you know that each will require about two bushels a day, just topped and tossed into a crib; and therefore you know how much of your Swedish turnip crop you ought to reserve for this purpose.

The grass, and even the lucerne, does not come into use, generally speaking, until the middle of May. You ought, therefore, according to the good English fashion of providing rather too much than too little, make a provision of Swedish turnips, so as to begin feeding upon them on the 1st of March, and ending the last week of May. Your oxen, fed upon these, will trip along as quickly as almost any horses, and will keep up their flesh during the whole of the sowing season; and, if circumstances should render it advisable, will be fat for the butcher by the next Christmas. SIXTH: the first cost of an ox or steer, three years old, would at this time, if of the North Devon or Sussex breed, be, perhaps, if in good condition, and ready broken into his work, from *fourteen* to *seventeen pounds*. You cannot have a horse of the same age under double the sum, if in good condition, and fit to do the work that an ox will perform. Every day of his life, until he be seven years old, the ox gets better and better; and, every day of his life the horse gets worse and worse, comes nearer and nearer to the day when his body, skin and all, must be sold for twenty or thirty shillings; and, during the four years which I have here supposed an ox might be advantageously kept at work, the horse will have cost, on an average, an additional five pounds at the least, for the blacksmith and the farrier, to say nothing here about the great difference in the four years' costs of food. At last, if from the age of the ox, or from any other circum-stance, it be desirable to fat him, he may bring you one-third more than his first cost, if not double the amount of it. In the case of selling off the stock of a farm, horses are a mere drug, if they be old or out of condition; but an ox will, either for working or fatting, always sell for his worth. He is something to be eaten, and has an intrinsic value, not at all depending upon adventitious circumstances, or upon taste or opinion. There is no question as to his soundness, no *warranty;* no roguish jockey here to come into play. There is one thing indeed, which the mention of the "warranty," here reminds me of; and that is, the dreadful falling off, which the general use of oxen would occasion in the practice of the Courts of Law. Our ears are constantly filled with the accounts of *horse-causes;* but, whoever heard of an *ox-cause?* The ox is the natural assistant of man in the labours of the field. So he was in the days of Moses, and throughout the whole of the periods of the transactions of which the Bible is a history. We read in the Bible of war-horses; of horses drawing chariots; but we never find an allusion to horses employed in the tillage of the land; for which, by their gentleness, by the nature of the food which they require, by their great docility, oxen seem to have been formed by nature. When I was in Long Island, I had a pair of large oxen and a pair of small ones; and, from that time I have been astonished at their not being more in use in England. If you want to do a very long day's work in summer time, it is necessary to rest in the middle of the day, and particularly if the weather be hot. What a clutter there is with horses in this case. They must be brought into the stable, rubbed down, fed at manger, and taken out again to the field, be the distance what it may; an ox is uncollared or unyoked, turned into the nearest field which has no crop in it; and, perhaps you may let him loose in the field where you are at plough, and he there, either on the unploughed ground, or round the hedges, gets him a luncheon, and is ready for you when you come back. The docility of oxen is beyond belief to those who have not been in the habit of using them. My man in Long Island, used, in summer time, to go out with his yoke and his bows

just at break of day; that is to say, as soon as he could see the
oxen at fifty or sixty yards from him; for, there it is a great thing
to get the main of the work done before ten o'clock, and after
five, in order to avoid the burning heat of the day. He generally
found the oxen lying down, in which respect again they were
so much better than the dainty and capricious horse, which
will sometimes stand upon his legs, even for a week together.
As soon as the man got a sight of the oxen, for the space was
large, he used to call out *Haw, boys*. At the second call, somewhat more loud than the former, the oxen used to rise up and
look at him, and then look at one another. When he approached
them near enough for his words to be distinctly heard, he used
to call out, *Come under,* upon which the oxen began to walk off
slowly towards him. The next words were, *Come under,* I TELL
ye, pronounced in a very commanding and even angry tone,
upon which the oxen set off to him at full trot, bringing their
heads up close to his body; and putting the yokes round their
necks, each fastened at the top with a little piece of wood,
away he walked, and they after him, into the field, where a
single plough-chain hooked on to a ring in the yoke, sent the
plough along in a minute. There are two objections stated
to the use of oxen. It is said that they go *slowly;* and so they
ought, and, on the finest arable farms that I ever saw, and I
believe are the finest in the world, I mean, in the vales of
Wiltshire, the horses go as slowly as foot can fall. It is the
history of the tortoise and the hare; the movements must be
slow in such a case; and, if the *time* be well husbanded, slow
movements are the best. The other objection is, that their
feet, unless they be shod, (which they never should be, by
me) disqualify them from travelling upon hard roads. I am not
recommending them as fit for road wagons, or vans or stage
coaches. I am recommending them for work upon a farm, which
includes going to the mill, going to coppice cart, going to
timber cart; as far as the outsides of the farm; all which work
I have seen them perform in the most excellent manner, at
Farmer Brazier's, at Worth, in Sussex, who always has a team

of four oxen, who has a pair of young ones always coming on, and who now and then fats a pair of the oldest, or sells them for the purpose of being fatted. A few miles upon even a hard turnpike road does the feet of an ox no harm; or else, how do those which are loaded with fat walk, sometimes a hundred and fifty miles, for the purpose of being devoured by the tax-eaters of London. These oxen are sometimes lame; but you will see whole droves of fat Devonshire oxen, without a lean one amongst them. An ox will go in a cart or wagon, just as well as a horse, and with more docility. The exceeding troublesomeness of relying solely upon the spade, and the necessity of my being constantly present here; besides the absolute necessity of a good deal of carting; made me, some-time ago in conversation with a friend, express my deep regret at not being able to bring over a Yankee with a pair of oxen, his yoke, and his bows. As it happened this friend had been in America himself, and was now using oxen in England, and kindly offered to send me two pair, ready broken to their work. I have them, and very docile and excellent oxen they are. I was afraid of being compelled to resort to the horse, the "head" carter, the well-informed blacksmith, and the scientific farrier. I trembled at the very thought, and was happily relieved from my distress by the intervention of this brother emigrant to America. For heavy or deep ploughing we take the two pair. For light ploughing, such as the inter-tillage above mentioned, one pair will be enough. These oxen cart the dung, bring in the corn, the mangel-wurzel roots and leaves, and, in short, do all the work upon the farm. I have one little horse to send into town and for other purposes; but an ox does very well to draw a cart to Kensington or to go any short distance, and, if not very frequently, over the pavement as far as Hyde Park Corner, or even further. About *farms*, however, there are, thank God, no pavements yet; and, therefore, it does appear to me to be the strangest thing in the world, that the use of oxen is not preferred to that of horses. The reason is, I believe, that farmers generally think that horses do more work than oxen; and another reason certainly is, that they will bear hurrying and whipping along, which oxen will not. An additional reason is, that horses can be sent with a load upon any hard roads, and that, generally speaking, oxen cannot; but, I want to know what great distances a farmer can send his team to. To market is the utmost. The market is not, on an average, ten miles distant from the farm; and that distance is not necessary to be gone over a great many times in the year, except the farm be very large, and then there must be several teams of oxen, amongst which this road work would of course be divided. In these objections, I therefore see nothing of any weight; and there remains only to speak of the *breaking-in*. In the first place, oxen are certainly more troublesome than cart horses to break into their work; but, with an inflexible resolution on the part of the master to see patience exercised, and all violence and rough and loud language avoided, and with attention to it himself, the breaking-in of a young ox is a matter of very little difficulty. In the first place, he should be kept for some time in company with the working oxen when they are not at work; he should feed them in the same pasture, or out of the same cribs in the yard. Then he should be tied up in a stall, for a month or so, every night by the side of one of the working oxen, and always the same ox, and on the same side of him. Then the collar (if collar be used) should be put on him, and he might wear it constantly. By-and-bye the rest of the harness. Then, he should be led out with the harness on, walking still on the same side of the same working ox. If yokes be used, he should be led yoked with the other ox without any thing to draw. The plough is the great doctor for horses as well as for oxen; for here there is no room, and no danger of injury to any thing. When first put to plough, the ground should be rather light, and the young ox should hardly feel that he has any thing behind him. If, however, he be stubborn, there should be no blows, and no loud scolding. Stop, pat him, and pat the other ox, and he will presently move on again and pull a little. If he lie down, which he sometimes will, let him lie till he be tired, and when he chooses to get up treat him very gently, as if he had been doing every thing that was right. By these means a young ox is broken to his labour in the course of a few days; and when that is once done, it is done for all the working part of his life. Rich pasture will not make him restive, as high feeding does a horse. With gentle treatment he is always of the same temper, is always of the same aptitude to labour: he is stronger with good feed than with bad; but no feed makes him unwilling to perform what you want him to perform. In conclusion, there is a question which is best, collars or yokes? in some respects one is, and in some respects the other; especially if you want an ox to go in shafts, to which the yoke cannot be easily applied. The collar, however, is not the natural thing for the ox, whose seats of strength are his neck and his horns, and not his shoulders. In America yokes are universal, and yokes prevail in Sussex. They are the least expensive, but if yokes be used with carts or wagons, there must be poles and not shafts; for the yoke is for two oxen and not for one. Mr Tull used a bull with a single yoke, a description of which is given in my *Year's Residence in America,* one of which I had there, and used it as long as I remained there; and such a yoke might be useful to some persons who persist in the narrow tillage.

WHEAT GROWING
On The Principles Laid Down by Jethro Tull

Jethro Tull (1674-1741), famous agricultural writer and farmer, was the inventor of row cultivation. He pioneered the system, which Cobbett describes, of sowing crops in rows wide enough apart to allow ploughing in between. Such ploughing he claimed did away with the need for manure. Tull published his book Horse Hoeing Husbandry *in 1731. Cobbett edited a new edition in 1822 and wrote an introduction of which this passage is a part.*

Mr. Tull's main principle is this, that *tillage will supply the place of manure;* and, his own experience shows, that, *a good* crop of wheat, for any number of years, may be grown, every year, upon the same land, without any manure, from first to last. The recent experiments that I have to mention confirm this; but, there is a difficulty in the case of *wheat,* mentioned by Mr. Tull himself, and which my experience also confirms.

It is naturally asked: "If perpetual crops of wheat can be produced on the same land without manure, *why is it not done?"* The answer to this question will be found in the following account of experiments.

Experiment in *Spring Wheat, in 1813*:

—On the 5th of April I sowed a land *broad-cast,* all across a field, which land contained thirteen *rod* or *perch* of ground. The land was in a very good state and very suitable to the crop. On each side of this land, and close to it, were made three ridges across the field at three feet asunder; and on these ridges was sown wheat, of the same sort, in a single drill, or row, upon the top of each ridge. So that there were two thirteen rods of this Tullian wheat. The broad-cast wheat was weeded and kept in very good order; and the Tullian was very nicely ploughed between and kept clear from weeds in the rows themselves.

Many persons saw this wheat while growing; and, though all were struck with the superior strength of straw and length and size of ear in the Tullian wheat, it appeared to everyone impossible that a land, nine feet wide, having only *three rows* of wheat on it, should bear as much as a like quantity of land *covered all over with wheat plants.* One of the gentlemen who saw the wheat, while growing, was Mr. Missing, the Barrister, of Posbrook, about seven miles from Botley. In order to *try the question,* as the lawyers call it, he and I made a bet as to comparative quantity; he on the broad-cast, and I upon the three rows.

The whole was reaped on the 16th of August, and housed on the 20th of the same month. In a little while after, it was, in presence of Mr. Missing, threshed out; the broad-cast by itself, and the two sets of three rows together. The two parcels were then winnowed. Then the Tullian parcel was divided into two; and the result was as follows:—

	Bushels	Gallons
Clean Wheat on Thirteen Perches in the Tullian manner, single rows on three-feet ridges	2	$5\frac{3}{4}$
Clean Wheat on Thirteen perches, Broad-cast	2	$7\frac{1}{2}$

The first was at the rate of 33 bushels 6 gallons to the acre, the last at the rate of 36 bushels half a gallon to the acre. I proposed, that, in estimating the crop, we should take the *seed* into view, and that my side should have *added* to it a quantity equal to the difference in the quantity of seed that the broad-cast had demanded more than the Tullian.

And, though my antagonist was by no means prepared to admit of any such interpretation of the terms of the bet, the reader will see, that, in estimating the clear relative produce, the *seed* must be taken into view. Now, then, the broad-cast was sown at the rate of three bushels to the acre; the Tullian at the rate of five gallons to the acre. The accounts, therefore, stand thus:

				Bushels	Gallons
BROAD-CAST	Produce	..	..	36	$0\frac{1}{2}$
	Deduct seed	..		3	0
		Clear Crop		33	$0\frac{1}{2}$
TULLIAN	Produce	..	..	33	6
	Deduct seed	..		0	5
		Clear Crop		33	1

Thus, then, the clear produce is rather greater in the Tullian than in the other.

Besides this experiment, I made another upon a part of one of the rows. I took ten feet in length from one of the rows, and thinned out the plants so as to leave only twelve plants in a foot in length. The produce of these ten feet was $1\frac{1}{2}$ pint within a mere trifle; and it weighed 1 lb. 5 oz. avoirdupois. Another ten feet, adjoining the former ten feet, but *not thinned out*, produced *not so much* by a wine-glass full; but it *weighed* precisely the same owing to the *unripeness* of the former, some of the ears of which were *quite green* when cut. Now observe; this thinned wheat, which would not require more than about a *gallon and a quarter* of seed to an acre produced at the rate of thirty-one bushels and nearly a half to the acre; and the wheat (*Spring* wheat too) weighed at the rate of 56 lbs. to the Winchester bushel. The weight of the *straw*, when threshed, on the ten feet, was 2 lbs. 7 oz. and, on an acre, it would have been 1 ton. 11 cwt. 2 grs. and 21 lbs.

It must be observed, however, that the year 1813 was the finest year for wheat within the memory of man. If the summer had been *cold,* and the harvest *backward,* the result, in the above case, would have been very different. The reader will note well what I have said about the *unripeness* of the ears on the ten feet of thinned wheat. Here the wheat was *backwarder* than in the rest of the rows which were *not thinned;* and, throughout the whole of the rows, the wheat was *backwarder* than in the broad-cast. This is the obstacle, and the only obstacle, to growing Tullian Wheat. The tillage makes the plants keep on growing to a much *later period* than when they stand thickly all over the ground, and have no tillage while growing. This *late growth,* and the *juiciness* of the stalks and leaves, expose the plants to that sort of *blight,* which makes the straw *speckled,* and sometimes gives it *a dark hue* all over. And whenever this blight lays hold, the *grain is thin* and light. The Tullian grains are much *larger* than those of the broad-cast; but, in this chilly and backward climate, you must, on an average of years, expect this blight, unless you sow *very early*. This I found to be the case. My land did not permit me to sow early; and, though I had in 1814, on a five acre field, *four quarters to the acre* of white Essex wheat, sown on *four feet* ridges, a single row upon a ridge, I found so much blight, generally, that I was obliged to discontinue the system, as to wheat, though, with regard to Swedish turnips (I never raised any other) I found it so excellent.

If you can sow your wheat *in August*, five times out of six you may escape the blight in the straw. But even then you must not sow too *thin;* always bearing in mind, that the thinner the plants stand, the *later* the wheat is in getting ripe; and the greater the chance of blight. It is to be observed, that Mr. Tull's was a *high-country* farm. I went to see it last fall. It is on a bottom of chalk. It is *Saint Foin* land. The sowing always takes place *early;* and this by no means applies to low, deep, rich, or stiff lands; though, we are now about to see that the system, even as to wheat, *may* be applied to strong lands.

When in Herefordshire, last fall, I heard of a gentleman, Mr. Beaman, who had, for many years, raised wheat in the Tullian manner. I had not time to wait on that gentleman myself; but, I put on paper certain questions, which I requested Mr. Palmer of Bollitree Castle to put to Mr. Beaman, and then to send the questions back to me, together with the answers. This was done by Mr. Palmer, in the following words:

How many years has Mr. Beaman grown Tullian Wheat upon any one field, without the intervention of other crops? – Answer. *Six years.*

Has he manured for it? – Ans. *No.*

What has been the width of the ridges? – Ans. *Begun at twenty seven inches, then four feet and a half, and, lastly, six feet.*

What have been the number of rows on each ridge? – Ans. *One row on twenty-seven inches, two rows on four feet and a half, and three rows on six feet.*

What is the season of sowing? – Ans. *The earlier the better; should be up before Michaelmas.*

How many ploughings, and the seasons of the ploughings?
– Ans. *The oftener the better, cannot plough too much, nor too near at the first ploughings.*

What is the quantity of seed to an acre? – Ans. *From half a peck to two bushels; but this must depend upon the soil, and upon the season of drilling.*

What is the expense of the hand-hoeings? – Ans. *From eighteen-pence to two shillings an acre.*

What is the general amount of the crops? – Ans. *Twenty-three bushels to an acre.*

Does it blight *more* than other wheat? – Ans. *Thinks it does.*

Please to get samples of as many years as possible. – Ans. *The samples got are of the years 1820 and 1821.* – N.B. Very good wheat. I guess it to weigh 57 lbs. or 58 lbs. a bushel.

What sort, or sorts of wheat, has he drilled? – Ans. *Has drilled all sorts; prefers a short stiff straw.*

What is the quantity of straw? – Ans. *About a load and a half to an acre.*

To the above Mr. Palmer added, that Mr. Beaman calls his land, "a marly clay". "The top-soil", says Mr. Palmer, "is a stiff, dark-coloured loam, upon a marl and clay bottom". Mr. Beaman thinks a single plant in about 3 inches would grow the greatest crop of wheat; but, then, the wheat should *be up in August.*

I am very much pleased with this information from Mr. Beaman; because *there is the thing now to be seen*! Mr. Tull continued his wheat crops to the harvesting of the *twelfth* upon the same land without manure; and, when he *concluded* his work, he had, as he informs us in a Memorandum, the *thirteenth* crop coming on, likely to be very good. Mr. Beaman had had *six crops*, and the average of his crops had been *twenty three bushels to the acre*, which is more than Arthur Young allows as the average of the Kingdom. However, is there any farmer in England (no matter what his land may be) who would not be glad to get twenty-three bushels of wheat from an acre, year after year, *without manure* and with so little expense as that of four or five horse-hoeings and two shillings an acre hand-hoeing? The blight is the only obstacle; but, perhaps, Lammas Wheat sown *in August*, or Spring (blended) Wheat sown *early in March*, would five times out of six, escape the blight; and, unquestionably, a field of Tullian wheat in ear, with the ridges straight and the land clean, is the most beautiful thing in the vegetable world. It is not *grand* like the Indian corn; but, it is even more beautiful than that. After three or four crops, there is very little trouble from *weeds* or *grass*. The land is prepared for any crop; and it is bearing a good crop, while the preparation is going on.

Besides the *early sowing*, care must be taken not to sow *too thin*. By thick sowing along the drills, you get the plants to stave one another a little, and the wheat ripens at an *earlier period*. Mr. Tull, in his Memorandum, says, that he put *one row* on a ridge in part of a field, the last year of his farming; but missed ascertaining the effect, owing to his being ill and to the hurry of the harvest. I am inclined to think, that it would not have answered; for, the fault always is, that the plants *keep on growing to too late a season*; and the nearer you go to them with the plough, the more *gross* they will be and the later they will continue to grow. And, we see, that Mr. Beaman began with *one row* on a ridge; but, that he has now come to *three rows*.

This grossness produced by the tillage does, however, fully show the truth of the *grand principle* of Tull; namely, that *tillage*, and tillage alone, will create and supply the *food of plants*, and will, in many cases, *render manure wholly unnecessary*.

THE MILL

I am not visionary enough to want to supersede mills. I know that the wind and the water have been wisely called in to the assistance of man; but, it is to his *assistance* and not to his *starvation*. My mill is in my granary; at Kensington it was in a house in the garden. There are places about every gentleman's house, all round this town. The expense of the grinding, reckoning that you have the bran, is next to nothing at all; and,

two men, while they are at the work, are in the dry and are
well fed, instead of being shivering about the streets half
starved. I used to get my flour from a miller in the country who
is a very worthy man, and he charged me at the lowest price;
but still, by grinding the wheat, I have my bread considerably
cheaper than I could from his flour, which was of precisely the
same quality as my own. Any gentleman who has a mind to
try this, will soon find himself repaid by the sweetness of his
bread, that is to say, if he buy the best wheat, which it is always
economical to do. The mill is put up in the space of an hour, by
any handy carpenter. It is fixed against a post in the room. The
mill-maker or his men will attend, and show anybody, in a few
minutes, how to perform the whole thing, which, as I said before,
has no mystery in it. I have bought, this year* my wheat to
grind into flour. But, let any reader be pleased to reflect on
the monstrous folly of my sending wheat of my own growth
from Barn-Elm to Uxbridge market, that being the nearest
considerable corn-market; selling it there to some corn-dealer,
to a certainty he selling it again to some miller; he selling it
again to some Hammersmith meal-man, perchance; the meal-
man, after adding to the *waust improvements* that it had probably
undergone in the hands of the trading miller, selling it to some
baker at Hammersmith, in whose hands it might, possibly, not
escape other improvements as "waust" as the former; and then,
at the end of all this, suppose me sending to Hammersmith to
purchase bread thus "waustly improved", and carrying within
it, or, rather, making me leave behind it, the profits of these
four intervening men; suppose all this, and then you suppose
no more than what is generally the case in this country of
miserable paupers and insolent idlers. Oh, no; that which I
and my family eat in the shape of bread, shall be made into
that shape on the spot on which the wheat shall grow.
* 1827

MAKING BREAD

As to the act of making bread, it would be shocking indeed if that had to be taught by the means of books. Every woman, high or low, ought to know how to make bread. If she do not, she is unworthy of trust and confidence: and, indeed, a mere burden upon the community. Yet it is but too true, that many women, even amongst those who have to get their living by their labour, know nothing of the making of bread, and seem to understand little more about it than the part which belongs to its consumption. A Frenchman, Mr Cusar, who had been born in the West Indies, told me, that till he came to Long Island he never knew how the flour came; that he was surprised when he learnt that it was squeezed out of little grains that grew at the tops of straw; for that he had always had an idea that it was got out of some large substance like the yams that grow in tropical climates. He was a very sincere and good man, and I am sure he told me truth. And this may be the more readily believed when we see so many women in England who seem to know no more of the constituent parts of a loaf than they know of those of the moon. Servant women in abundance appear to think that loaves are made by the maker, as knights are made by the king; things of their pure creation, a creation too in which no one else can participate. Now is not this an enormous evil? And whence does it come? Servant women are the children of the labouring classes; and they would all know how to make bread, and know well how to make it too, if they had been fed on bread of their mothers' and their own making.

In the last Number, I observed that I hoped it was unnecessary for me to give any directions as to the mere act of making bread. But several correspondents inform me that without these directions a conviction of the utility of baking bread at home is of no use to them. Therefore I shall here give those directions, receiving my instructions here from one who, I thank God, does know how to perform this act.

Suppose the quantity be a bushel of flour. Put this flour into a trough that people have for the purpose, or it may be in a clean smooth tub of any shape if not too deep, and if sufficiently large. Make a pretty deep hole in the middle of this heap of

flour. Take (for a bushel) a pint of good fresh yeast, mix it and
stir it well up in a pint of soft water milk-warm. Pour this into
the hole in the heap of flour. Then take a spoon and work it
round the outside of this body of moisture so as to bring into
that body, by degrees, flour enough to make it form a thin
batter, which you must stir about well for a minute or two.
Then take a handful of flour and scatter it thinly over the head
of this batter, so as to hide it. Then cover the whole with a
cloth to keep it warm; and this covering, as well as the situation
of the trough, as to distance from the fire, must depend on the
nature of the place and state of the weather as to heat and cold.
When you perceive that the batter has risen enough to make
cracks in the flour that you covered it over with, you begin to
form the whole mass into dough, thus; you begin round the
hole containing the batter, working the flour into the batter,
and pouring in, as it is wanted to make the flour mix with the
batter, soft water milk-warm, or milk, as hereafter to be
mentioned. Before you begin this you scatter the salt over the
heap at the rate of half a pound to a bushel of flour. When you
have got the whole sufficiently moist, you knead it well. This
is a grand part of the business; for, unless the dough be well
worked, there will be little round lumps of flour in the loaves;
and besides, the original batter, which is to give fermentation
to the whole, will not be duly mixed. The dough must, therefore be well worked. The fists must go heartily into it. It must
be rolled over, pressed out, folded up and pressed out again,
until it be completely mixed and formed into a stiff and tough
dough. This is labour, mind. I have never quite liked baker's
bread since I saw a great heavy fellow in a bakehouse in
France, kneading bread with his naked feet! His feet looked
very white, to be sure; whether they were of that colour before
he got into the trough I could not tell. God forbid that I should
suspect that this is ever done in England! It is labour, but what
is exercise other than labour? Let a young woman bake a
bushel once a week, and she will do very well without phials
and gallipots.

Thus, then the dough is made, and when made it is to be formed into a lump in the middle of the trough, and with a little dry flour thinly scattered over it, covered over again to be kept warm and to ferment; and in this state, if all be done rightly, it will not have to remain more than about 15 or 20 minutes.

In the meanwhile the oven is to be heated, and this is much more than half the art of the operation. When an oven is properly heated, can be known only by actual observation. Women who understand the matter know when the heat is right the moment they put their faces within a yard of the oven-mouth; and once or twice observing is enough for any person of common capacity. But this much may be said in the way of rule, that the fuel (I am supposing a brick oven) should be dry (not rotten) wood, and not mere brushwood, but rather fagot-sticks. If larger wood, it ought to be split up into sticks not more than two or two and a half inches through. Brushwood that is strong, not green and not cold, if it be hard in its nature and has some sticks in it, may do. The woody parts of furze, or ling, will heat an oven very well. But the thing is, to have a lively and yet somewhat strong fire, so that the oven may be heated in about 15 minutes, and retain its heat sufficiently long.

The oven should be hot by the time that the dough has remained in the lump about 20 minutes. When both are ready, take out the fire, and wipe the oven out clean, and at nearly about the same moment, take the dough out upon the lid of the baking trough, or some proper place, cut it up into pieces, and make it up into loaves, kneading it again into the separate parcels; and as you go on, shaking a little flour over your board, to prevent the dough from adhering to it. The loaves should be put into the oven as quickly as possible after they are formed; when in, the oven lid or door should be fastened up very closely; and if all be properly managed, loaves of about the size of quartern loaves will be sufficiently baked in about two hours. But they usually take down the lid and look at the bread, in order to see how it is going on.

And what is there worthy of the name of plague or trouble in all this? Here is no dirt, no filth, no rubbish, no litter, no slop. And pray what can be pleasanter to behold? Give me for a beautiful sight a neat and smart woman, heating her oven and setting her bread! And if the bustle does make the sign of labour glisten on her brow, where is the man that would not kiss that off, rather than lick the plaster from the cheek of a duchess?

And what is the result? Why, good, wholesome food, sufficient for a considerable family for a week, prepared in about three or four hours. To get this quantity of food, fit to be eaten, in the shape of potatoes, how many fires! what a washing, what a boiling, what a peeling, what a slopping, and what a messing! The cottage everlastingly in a litter; the woman's hands everlastingly wet and dirty; the children grimed up to the eyes with dust fixed on by potatoe starch; and ragged as colts, the poor mother's time all being devoted to the everlasting boiling of the pot! Can any man, who knows anything of the labourer's life, deny this? And will, then, anybody, except the old shuffle-breeches band of the *Quarterly Review*, who have all their lives been moving from garret to garret, who have seldom seen the sun, and never the dew except in print; will anybody, except these men, say, that the people ought to be taught to use potatoes as a substitute for bread?

KEEPING PIGS

As in the case of cows so in that of pigs, much must depend upon the situation of the cottage; because all pigs will graze; and therefore, on the skirts of forests or commons, a couple or three pigs may be kept, if the family be considerable; and especially if the cottager brew his own beer, which will give grains to assist the wash. Even in lanes, or on the sides of great roads, a pig will find a good part of his food from May to November; and if he be yoked, the occupiers of the neighbourhood must be churlish and brutish indeed, if they give the owner any annoyance.

Let me break off here for a moment to point out to my readers the truly excellent conduct of Lord Winchelsea and Lord Stanhope, who, as I read, have taken great pains to make the labourers on their estates comfortable, by allotting to each a piece of ground sufficient for the keeping of a sow. I once, when I lived at Botley, proposed to the copy-holders and other farmers in my neighbourhood that we should petition the Bishop of Winchester, who was lord of the manors thereabouts, to grant titles to all the numerous persons called trespassers on the wastes; and also to give titles to others of the poor parishioners who were willing to make, on the skirts of the wastes, enclosures not exceeding an acre each. This I am convinced would have done a great deal towards relieving the parishes, then greatly burdened by men out of work. This would have been better than digging holes one day to fill them up the next. Not a single man would agree to my proposal! One, a bull-frog farmer (now, I hear, pretty well sweated down), said it would only make them saucy! And one, a true disciple of Malthus, said, that to facilitate their rearing children was a harm! This man had at the time, in his own occupation, land that had formerly been six farms, and he had, too, ten or a dozen children. I will not mention names; but this farmer will now, perhaps, have occasion to call to mind what I told him on that day, when his opposition, and particularly the ground of it, gave me the more pain, as he was a very industrious, civil, and honest man. Never was there a greater mistake than to suppose that men are made saucy and idle by just and kind treatment. Slaves are always lazy and saucy; nothing but the lash will extort from them either labour or respectful deportment. I never met with a saucy Yankee (New Englander) in my life. Never servile, always civil. This must necessarily be the character of freemen living in a state of competence. They have nobody to envy, nobody to complain of, they are in good humour with mankind. It must, however, be confessed that very little, comparatively speaking, is to be accomplished by the individual efforts even of benevolent men like the two noblemen before mentioned. They have a strife to maintain against the general tendency of the national state of things. It is by general and indirect means, and not by partial and direct and positive regulations, that so great a good as that which they generously aim at can be accomplished. When we are to see such means adopted God only knows; but if much longer delayed, I am of opinion that they will come too late to prevent something very much resembling a dissolution of society.

The cottager's pig should be bought in the spring, or late in winter; and being then four months old, he will be a year old before killing time; for it should always be borne in mind, that this age is required in order to ensure the greatest quantity of meat from a given quantity of food. If a hog be more than a year old, he is the better for it. The flesh is more solid and more nutritious than that of a young hog, much in the same degree that the mutton of a full-mouthed wether is better than that of a younger wether. The pork or bacon of young hogs, even if fatted with corn, is very apt to boil out, as they call it; that is to say, come out of the pot smaller in bulk than it goes in. When you begin to fat, do it by degrees, especially in the case of hogs under a year old. If you feed high all at once, the hog is apt to surfeit, and then a great loss of food takes place. Peas, or barley-meal, is the food; the latter rather the best, and does the work quicker. Make him quite fat by all means. The last bushel, even if he sit as he eat, is the most profitable. If he can walk two hundred yards at a time, he is not well fatted. Lean bacon is the most wasteful thing that any family can use. In short, it is uneatable, except by drunkards, who want something to stimulate their sickly appetite. The man who cannot live on solid fat bacon, well fed and well cured, wants the sweet sauce of labour, or is fit for the hospital. But, then, it must be bacon, the effect of barley or peas (not beans), and not of whey, potatoes, or messes of any kind. It is frequently said, and I know that even farmers say it, that bacon made from corn costs more than it is worth! Why do they take care to have it then? They know better. They know well that it is the very cheapest they can have; and they who look at both ends and both sides of every cost would as soon think of shooting their hogs as of fatting them on messes; that is to say, for their own use, however willing they might now-and-then be to regale the Londoners with a bit of potato-pork.

About Christmas, if the weather be coldish, is a good time to kill. If the weather be very mild, you may wait a little longer, for the hog cannot be too fat. The day before killing he should have no food. To kill a hog nicely is so much of a profession that it is better to pay a shilling for having it done, than to stab and hack and tear the carcass about. I shall not speak of pork, for I would by no means recommend it. There are two ways of going to work to make bacon; in one you take off the hair by scalding. This is the practice in most parts of England, and all over America. But the Hampshire way, and the best way, is to burn the hair off. There is a great deal of difference in the consequences. The first method slackens the skin, opens all the pores of it, makes it loose and flabby by drawing out the roots of the hair. The second tightens the skin in every part, contracts all the sinews and veins in the skin, makes the flitch a solider thing, and the skin a better protection to the meat. The taste of the meat is very different from that of a scalded hog, and to this chiefly it was that Hampshire bacon owed its reputation of excellence. As the hair is to be burnt off it must be dry, and care must be taken that the hog be kept on dry litter of some sort the day previous to killing. When killed he is laid upon a narrow bed of straw, not wider than his carcass, and only two or three inches thick. He is then covered all over thinly with straw, to which, according as the wind may be, the fire is put at one end. As the straw burns, it burns the hair. It requires two or three coverings and burnings, and care is taken that the skin be not in any part burnt or parched. When the hair is all burnt off close, the hog is scraped clean, but never touched with water. The upper side being finished the hog is turned over, and the other side is treated in like manner. This work should always be done before day-light, for in the day-light you cannot so nicely discover whether the hair be sufficiently burnt off. The light of the fire is weakened by that of the day. Besides, it makes the boys get up very early for once at any rate, and that is something, for boys always like a bonfire.

The inwards are next taken out, and if the wife be not a slattern, here, in the mere offal, in the mere garbage, there is food, and delicate food, too, for a large family for a week; the hog's puddings for the children, and some for neighbours' children, who come to play with them; for these things are by no means to be overlooked, seeing that they tend to the keeping alive of that affection in children for their parents, which later in life will be found absolutely necessary to give effect to wholesome precept, especially when opposed to the boisterous passions of youth.

The butcher the next day cuts the hog up, and then the house is filled with meat! Souse, griskins, blade-bones, thigh bones, spare-ribs, chines, belly-pieces, cheeks, all coming into use one after the other, and the last of the latter not before the end of about four or five weeks. But about this time it is more than possible that the Methodist parson will pay you a visit. It is remarked in America, that these gentry are attracted by the squeaking of the pigs, as the fox is by the cackling of the hen. This may be called slander, but I will tell you what I did know to happen. A good honest careful fellow had a spare-rib, on which he intended to sup with his family after a long and hard day's work at coppice-cutting. Home he came at dark with his two little boys, each with a nitch of wood that they had carried four miles, cheered with the thought of the repast that awaited them. In he went, found his wife, the Methodist parson, and the whole troop of the sister-hood, engaged in prayer, and on the table lay scattered the clean-polished bones of the spare-rib! Can any reasonable creature believe, that, to save the soul, God requires us to give up the food necessary to sustain the body? Did Saint Paul preach this? He who, while he spread the gospel abroad, worked himself, in order to have it to give to those who were unable to work? Upon what, then, do these modern saints, these evangelical gentlemen, found their claim to live on the labour of others?

All the other parts taken away, the two sides that remain, and that are called flitches, are to be cured for bacon. They are first rubbed with salt on their insides, or flesh sides, then placed one on the other, the flesh sides uppermost, in a salting trough which has a gutter round its edges to drain away the brine; for to have sweet and fine bacon the flitches must not lie sopping in brine, which gives it that sort of taste which barrel-pork and sea-jonk have, and than which nothing is more villainous. Everyone knows how different is the taste of fresh dry salt from that of salt in a dissolved state. The one is savoury, the other nauseous. Therefore, change the salt often. Once in four or five days. Let it melt and sink in but let it not lie too long. Change the flitches. Put at the bottom that which was first put on the top. Do this a couple of times. This mode will cost you a great deal more in salt, or rather in taxes, than the sopping mode; but without it your bacon will not be sweet and fine, and will not keep so well. As to the time required for making the flitches sufficiently salt, it depends on circumstances; the thickness of the flitch, the state of the weather, the place wherein the salting is going on. It takes a longer time for a thick than for a thin flitch; it takes longer in dry than in damp weather; it takes longer in a dry than in a damp place. But for the flitches of a hog of twelve score, in weather not very dry or very damp, about six weeks may do; and as yours is to be fat, which receives little injury from over-salting, give time enough; for you are to have bacon till Christmas comes again. The place for salting should, like a dairy, always be cool, but always admit of a free circulation of air; confined air, though cool, will taint meat sooner than the mid-day sun accompanied with a breeze. Ice will not melt in the hottest sun so soon as in a close and damp cellar. Put a lump of ice in cold water, and one of the same size before a hot fire, and the former will dissolve in half the time that the latter will. Let me take this occasion of observing, that an ice-house should never be under ground, or under the shade of trees. That the bed of it ought to be three feet above the level of the ground, that this bed ought to consist of something that will admit the drippings to go instantly off, and the house should stand in a place open to the sun and air. This is the way they have the ice-houses under the burning sun of Virginia; and here they keep their fish and meat as fresh and sweet as in winter, when at the same time neither will keep for twelve hours, though let down to the depth of a hundred feet in a well. A Virginian, with some poles and straw, will stick up an ice-house for ten dollars, worth a dozen of those ice-houses, each of which costs our men of taste as many scores of pounds. It is very hard to imagine, indeed, what any one should want ice for, in a country like this, except for clodpole boys to slide upon, and to drown cockneys in skating-time; but if people must have ice in summer, they may as well go a right way as a wrong way to get it.

However, the patient that I have at this time under my hands wants nothing to cool his blood, but something to warm it, and therefore I will get back to the flitches of bacon, which are now to be smoked; for smoking is a great deal better than merely drying; as is the fashion in the dairy-counties in the West of England. When there were plenty of farm-houses, there were plenty of places to smoke bacon in; since farmers have lived in gentlemen's houses, and the main part of the farm-houses have been knocked down, these places are not so plenty. However, there is scarcely any neighbourhood without a chimney left to hang bacon up in. Two precautions are necessary; first, to hang the flitches where no rain comes down upon them; second, not to let them be so near the fire as to melt. These precautions taken, the next is, that the smoke must proceed from wood, not turf, peat or coal. Stubble or litter might do: but the trouble would be great. Fir or deal smoke is not fit for the purpose. I take it, that the absence of wood, as fuel, in the dairy-counties, and in the North, has led to the making of pork and dried bacon. As to the time that it requires to smoke a flitch, it must depend a good deal upon whether there be a constant fire beneath, and whether the fire be large or small. A month may do, if the fire be pretty constant, and such as a farmhouse fire usually is. But oversmoking, or rather, too long hanging in the air, makes the bacon rust. Great attention should, therefore, be paid to this matter. The flitch ought not to be dried up to the hardness of a board, and yet it ought to be perfectly dry. Before you hang it up, lay it on the floor, scatter the flesh side pretty thickly over with bran, or with some fine sawdust other than that of deal or fir. Rub it on the flesh, or pat it well down upon it. This keeps the smoke from getting into the little openings, and makes a sort of crust to be dried on; and, in short, keeps the flesh cleaner than it would otherwise be.

To keep the bacon sweet and good, and free from nasty things that they call hoppers; that is to say a sort of skipping maggots, engendered by a fly which has a great relish for bacon; to provide against this mischief, and also to keep the bacon from becoming rusty, the Americans, whose country is so hot in summer, have two methods. They smoke no part of the hog except the hams, or gammons. They cover these with coarse linen clothes, such as the finest hop-bags are made of, which they sew neatly on. They then white-wash the cloth all over with lime white-wash, such as we put on walls, their lime being excellent stone lime. They give the ham four or five washings, the one succeeding as the former gets dry; and in the sun, all these washings are put on in a few hours. The flies cannot get through this; and thus the meat is preserved from them. The other mode, and that is the mode for you, is to sift fine some clean and dry wood-ashes. Put some at the bottom of a box or chest, which is long enough to hold a flitch of bacon. Lay in one flitch; and then cover this with six or eight inches of the ashes. This will effectually keep away all flies; and will keep the bacon as fresh and good as when it came out of the chimney, which it will not be for any great length of time, if put on a rack, or kept hung up in the open air. Dust, or even sand, very, very dry, would, perhaps, do as well. The object is not only to keep out the flies, but the air. The place where the chest, or box, is kept, ought to be dry; and, if the ashes should get damp (as they are apt to do from the salts they contain), they should be put in the fire-place to dry, and then be put back again. Peat-ashes, or turf-ashes, might do very well for this purpose. With these precautions, the bacon will be as good at the end of the year as on the first day; and it will keep two, or even three years, perfectly good, for which, however, there can be no necessity.

Now, then, this hog is altogether a capital thing. The other parts will be meat for about four or five weeks. The lard, nicely put down, will last a long while for all the purposes for which it is wanted. To make it keep well there should be some salt put into it. Country children are badly brought up if they do not like sweet lard spread upon bread, as we spread butter. Many a score hunches of this sort have I eaten, and I never knew what poverty was. I have eaten it for luncheon, at the houses of good substantial farmers in France and Flanders. I am not now frequently so hungry as I used to be; but I should think it no hardship to eat sweet lard instead of butter. But, now-a-days, the labourers, and especially the female part of them have fallen into the taste of niceness in food and finery in dress; a quarter of a bellyful and rags are the consequence. The food of their choice is high-priced, so that, for the greater part of their time, they are half-starved. The dress of their choice is showy and flimsy, so that, today, they are ladies, and tomorrow ragged as sheep with the scab. But has not Nature made the country girls as pretty as ladies? Oh yes (bless their rosy cheeks and white teeth!) and a great deal prettier, too! But are they less pretty, when their dress is plain and substantial, and when the natural presumption is, that they have smocks as well as gowns, than they are when drawn off in the frail fabric of Sir

Robert Peel, 'where tawdry colours strive with dirty white,' exciting a violent suspicion that all is not as it ought to be nearer the skin, and calling up a train of ideas extremely hostile to that sort of feeling which every lass innocently and commendably wishes to awaken in her male beholders? Are they prettiest when they come through the wet and dirt safe and neat; or when their draggled dress is plastered to their backs by a shower of rain? However, the fault has not been theirs, nor that of their parents. It is the system of managing the affairs of the nation. This system has made all flashy and false, and has put all things out of their place. Pomposity, bombast, hyperbole, redundancy and obscurity, both in speaking and in writing; mock-delicacy in manners, mock-liberality, mock-humanity and mock-religion. Pitt's false money, Peel's flimsy dresses, Wilberforce's potatoe diet, Castlereagh's and Mackintosh's oratory, Walter Scott's poems, Walter's and Stoddart's paragraphs, with all the bad taste and baseness and hypocrisy which they spread over this country; all have arisen, grown, branched out, bloomed and borne together; and we are now beginning to taste of their fruit. But, as the fat of the adder is, as is said, the antidote to its sting; so in the Son of the Great Worker of Spinning Jennies, we have, thanks to the Proctors and Doctors of Oxford, the author of that Bill, before which this false, this flashy, this flimsy, this rotten system will dissolve as one of his father's pasted calicoes does at the sight of the washing-tub!

'What,' says the Cottager, 'has all this to do with hogs and bacon?' Not directly with hogs and bacon, indeed; but it has a great deal to do, my good fellow, with your affairs, as I shall, probably, hereafter more fully show, though I shall now leave you to the enjoyment of your flitches of bacon, which, as I before observed, will do ten thousand times more than any Methodist parson, or any other parson (except, of course, those of our church), to make you happy, not only in this world but in the world to come. Meat in the house is a great source of harmony, a great preventer of the temptation to commit those

things which, from small beginnings, lead, finally, to the most
fatal and atrocious results; and I hold that doctrine to be truly
damnable which teaches that God has made any selection, any
condition relative to belief, which is to save from punishment
those who violate the principles of natural justice.

Some other meat you may have, but bacon is the great thing.
It is always ready: as good cold as hot; goes to the field or the
coppice conveniently; in harvest, and other busy times, demands the pot to be boiled only on a Sunday; has twice as
much strength in it as any other thing of the same weight; and
in short, has in it every quality that tends to make a labourer's
family able to work and be well off. One pound of bacon, such
as that which I have described, is, in a labourer's family, worth
four or five of ordinary mutton or beef, which are great part
bone, and which, in short, are gone in a moment. But always
observe, it is fat bacon that I am talking about. There will, in
spite of all that can be done, be some lean in the gammons,
though comparatively very little, and therefore you ought to
begin at that end of the flitches, for old lean bacon is not good.

Now, as to the cost. A pig (a spayed sow is best) bought in
March, four months old, can be had now for fifteen shillings.
The cost till fattening time is next to nothing to a Cottager;
and then the cost, at the present price of corn, would, for a hog
of twelve score, not exceed three pounds; in the whole four
pounds five; a pot of poison a week bought at the public-house
comes to twenty-six shillings of the money; and more than
three times the remainder is generally flung away upon the
miserable tea. I have, indeed, shown, that if the tea were
laid aside, the labourer might supply his family well with beer
all the year round, and have a fat hog of even fifteen score for
the cost of the tea, which does him, and can do him, no good
at all.

The feet, the cheeks, and other bone, being considered, the
bacon and lard, taken together, would not exceed sixpence a
pound. Irish bacon is 'cheaper'. Yes, lower-priced. But I will
engage that a pound of mine, when it comes out at the pot (to

say nothing of the taste) shall weigh as much as a pound and a half of Irish, or any dairy or slop-fed bacon, when that comes out of the pot. No, no; the farmers joke when they say that their bacon costs them more than they could buy bacon for. They know well what it is they are doing; and besides, they always forget, or rather, remember not to say, that the fatting of a large hog yields them three or four loads of dung, really worth more than ten or fifteen of common yard dung. In short, without hogs farming could not go on; and it never has gone on in any country in the world. The hogs are the great stay of the whole concern. They are much in small space; they make no show, as flocks and herds do; but without them cultivation of the land would be a poor, a miserably barren concern.

The reader will think, that I shall never cease talking about *hogs;* but, I have now done, only I will add, that, in keeping hog in a *growing state*, we must never forget their *lodging*! A few boards, flung carelessly over a couple of rails, and no litter beneath, is not the sort of bed for a hog. A place of suitable size, large rather than small, well sheltered on every side, covered with a roof that lets in no wet or snow. No opening, except a door-way big enough for a hog to go in; and the floor constantly well bedded with leaves of trees, *dry*, or, which is the best thing, and what a hog deserves, *plenty of clean straw*. When I make up my hogs' lodging place for winter, I look well at it, and consider, whether, upon a pinch, I could, for once and away, make shift to lodge in it myself. If I *shiver at the thought*, the place is not good enough for my hogs. It is not in the nature of a hog to sleep in the cold. Look at them. You will see them, if they have the means, *cover themselves over* for the night. This is what is done by neither horse, cow, sheep, dog or cat. And this should admonish us to provide hogs with warm and comfortable lodging. Their sagacity in providing against cold in the night, when they have it in their power to make such provision, is quite wonderful. You see them looking about for the warmest spot: then they go to work, raking up the litter

so as to break the wind off; and when they have done their best,
they lie down. I had a sow that had some pigs running about
with her in April last. There was a place open to her on each
side of the barn. One faced the east and the other the west;
and, I observed, that she sometimes took to one side and sometimes to the other. One evening her pigs had gone to bed on the
east side. She was out eating till it began to grow dusk. I saw
her go into her pigs, and was surprised to see her come out
again; and therefore, looked a little to see what she was after.
There was a high heap of dung in the front of the barn to the
south. She walked up to the top of it, raised her nose, turned it
very slowly, two or three times, from the north-east to the
north-west, and back again, and at last, it settled at about
south-east, for a little bit. She then came back, marched away
very hastily to her pigs, roused them up in a great bustle, and
away she trampled with them at her heels to the place on the
west side of the barn. There was so little wind, that I could not
tell which way it blew, till I took up some leaves, and tossed
them in the air. I then found, that it came from the precise
point at which her nose had settled at. And thus was I convinced, that she had come out to ascertain which way the wind
came, and finding it likely to make her young ones cold in the
night, she had gone and called them up, though it was nearly
dark, and taken them off to a more comfortable berth. Was
this an *instinctive,* or was it a *reasoning* proceeding? At any rate,
let us not treat such animals as if they were stocks and stones.

FOWLS

These are kept for two objects; their flesh and their eggs.
As to laying-hens, there are some means to be used to secure
the use of them in winter. They ought not to be old hens.
Pullets, that is, birds hatched in the foregoing spring, are,
perhaps, the best. At any rate, let them not be more than two
years old. They should be kept in a warm place, and not let
out, even in the day time, in wet weather; for one good sound
wetting will keep them back for a fortnight. The dry cold, even

in the severest cold, if dry, is less injurious than even a little wet in winter time. If the feathers get wet, in our climate, in winter, or in short days, they do not get dry for a long time; and this it is that spoils and kills many of our fowls.

The French, who are great egg eaters, take singular pains as to the food of laying-hens in winter. They let them out very little, even in their fine climate, and give them very stimulating food; barley boiled, and given them warm; curds, buck-wheat (which, I believe, is the best thing of all except curds); parsley and other herbs chopped fine; leeks chopped in the same way; also apples and pears chopped very fine; oats and wheat cribbled; and sometimes they give them hemp-seed, and the seed of nettles; or dried nettles harvested in summer, and boiled in the winter. Some give them ordinary food, and, once a day, toasted bread sopped in wine. White cabbages chopped up are very good in winter for all sorts of poultry.

This is taking a great deal of pains; but the produce is also great and very valuable in winter; for, as to preserved eggs, they are things to run from and not after. All this supposes, however, a proper hen-house, about which we, in England, take very little pains. The vermin, that is to say, the lice, that poultry breed, are the greatest annoyance. And as our wet climate furnishes them, for a great part of the year, with no dust by which to get rid of these vermin, we should be very careful about cleanliness in the hen-houses. Many a hen, when sitting, is compelled to quit her nest to get rid of the lice. They torment the young chickens. And, in short, are a great injury. The fowl-house should, therefore, be very often cleaned out; and sand, or fresh earth, should be thrown on the floor. The nest should not be on shelves, or anything fixed; but little flat baskets something like those that the gardeners have in the markets in London, and which they call sieves, should be placed against the sides of the house upon pieces of wood nailed up for the purpose. By this means the nests are kept perfectly clean, because the baskets are, when necessary, taken down, the hay thrown out, and the baskets washed; which cannot be done, if

the nest be made in anything forming a part of the building.
Besides this the roosts ought to be cleaned every week, and the
hay changed in the nests of laying-hens. It is good to fumigate
the house frequently by burning dry herbs, juniper wood, cedar
wood, or with brimstone; for nothing stands so much in need
of cleanliness as a fowl-house, in order to have fine fowls and
plenty of eggs.

The ailments of fowls are numerous, but they would seldom be
seen if the proper care were taken. It is useless to talk of remedies
in a case where you have complete power to prevent the evil.
If well fed, and left perfectly clean, fowls will seldom be sick;
and, as to old age, they never ought to be kept more than a
couple or three years; for they get to be good for little as layers,
and no teeth can face them as food.

It is, perhaps, seldom that fowls can be kept conveniently
about a cottage; but when they can, three, four, or half a
dozen hens to lay in winter, when the wife is at home the
greater part of the time, are worth attention. They would
require but little room, might be bought in November and sold

in April, and six of them, with proper care, might be made to clear every week the price of a gallon of flour. If the labour were great, I should not think of it; but it is none; and I am for neglecting nothing in the way of pains in order to ensure a hot dinner every day in winter, when the man comes home from work. As to the fatting of fowls, information can be of no use to those who live in a cottage all their lives; but it may be of some use to those who are born in cottages and go to have the care of poultry at richer persons' houses. Fowls should be put to fat about a fortnight before they are wanted to be killed. The best food is barley-meal wetted with milk, but not wetted too much. They should have clear water to drink, and it should be frequently changed. Crammed fowls are very nasty things: but 'barn-door' fowls, as they are called, are sometimes a great deal more nasty. *Barn*-door would, indeed, do exceedingly well; but it unfortunately happens that the stable is generally pretty near to the barn. And now let any gentleman who talks about sweet barn-door fowls, have one caught in the yard, where the stable is also. Let him have it brought in, killed, and the craw taken out and cut open. Then let him take a ball of horse-dung from the stable-door; and let his nose tell him how very small is the difference between the smell of the horse-dung and the smell of the craw of his fowl. In short, roast the fowl, and then pull aside the skin at the neck, put your nose to the place, and you will almost think you are at the stable-door. Hence the necessity of taking them away from the barn-door a fort-night, at least, before they are killed. We know very well that ducks that have been fed upon fish, either wild ducks or tame ducks, will scent a whole room, and drive out of it all those who have not got pretty good constitutions. It must be so. Solomon says that all flesh is grass; and those who know any-thing about beef, know the difference between the effect of the grass in Herefordshire and Lincolnshire, and the effect of turnips and oil cake. In America they always take the fowls from the farm-yard, and shut them up a fortnight or three weeks before they be killed. One thing, however, about fowls

ought always to be borne in mind. They are never good for
anything when they have attained their full growth, unless
they be capons or poullards. If the pullets be old enough to
have little eggs in them, they are not worth one farthing; and
as to the cocks of the same age, they are fit for nothing but to
make soup for soldiers on their march, and they ought to be
taken for that purpose.

PIGEONS

A few of these may be kept about any cottage, for they are kept
even in towns by labourers and artizans. They cause but little
trouble. They take care of their own young ones; and they do
not scratch, or do any other mischief in gardens. They want
feeding with tares, peas or small beans; and buck-wheat is
very good for them. To begin keeping them, they must not
have flown at large before you get them. You must keep them
for two or three days, shut into the place which is to be their
home; and then they may be let out, and will never leave you
as long as they can get proper food, and are undisturbed by
vermin, or unannoyed exceedingly by lice.

The common dove-house pigeons are the best to keep. They
breed oftenest, and feed their young ones best. They begin to
breed at about nine months old, and if well kept, they will give
you eight or nine pair in the year. Any little place; a shelf in
the cow shed; a board or two under the eaves of the house; or,
in short, any place under cover, even on the ground floor, they
will sit and hatch and breed up their young ones in.

It is not supposed that there could be much profit attached
to them; but they are of this use; they are very pretty creatures;
very interesting in their manners; they are an object to delight
children, and to give them the early habit of fondness for
animals and of setting a value on them, which, as I have often
had to observe before, is a very great thing. A considerable part
of all the property of a nation consists of animals. Of course a
proportionate part of the cares and labours of a people appertains to the breeding and bringing to perfection those animals;

and, if you consult your experience, you will find that a labourer is, generally speaking, of value in proportion as he is worthy of being intrusted with the care of animals. The most careless fellow cannot hurt a hedge or ditch; but to trust him with the team, or the flock, is another matter. And, mind, for the man to be trustworthy in this respect, the boy must have been in the habit of being kind and considerate towards animals; and nothing is so likely to give him that excellent habit as his seeing, from his very birth, animals taken great care of, and treated with great kindness, by his parents, and now-and-then having a little thing to call his own.

RABBITS

In this case, too, the chief use, perhaps, is to give children those habits of which I have been just speaking. Nevertheless, rabbits are really profitable. Three does and a buck will give you a rabbit to eat for every three days in the year, which is a much larger quantity of food than any man will get by spend-

ing half his time in the pursuit of wild animals, to say nothing of the toil, the tearing of clothes, and the danger of pursuing the latter.

Everybody knows how to knock up a rabbit hutch. The does should not be allowed to have more than seven litters in a year. Six young ones to a doe is all that ought to be kept; and then they will be fine. Abundant food is the main thing; and what is there that a rabbit will not eat? I know of nothing green that they will not eat; and if hard pushed, they will eat bark, and even wood. The best thing to feed the young ones on when taken from the mother, is the carrot, wild or garden, parsnips, Swedish turnips, roots of dandelion; for too much green or watery stuff is not good for weaning rabbits. They should remain as long as possible with the mother. They should have oats once a day; and, after a time, they may eat anything with safety. But if you give them too much green at first when they are weaned, they rot as sheep do. A variety of food is a great thing; and, surely, the fields and gardens and hedges furnish this variety! All sorts of grasses, strawberry leaves, ivy, dandelions, the hog-weed or wild parsnip, in root, stem and leaves. I have fed working horses six or eight in number, upon this plant for weeks together. It is a tall bold plant, that grows in prodigious quantities in the hedges and coppices in some parts of England. It is the perennial parsnip. It has flower and seed precisely like those of the parsnip; and hogs, cows and horses are equally fond of it. Many a half-starved pig have I seen within a few yards of cart-loads of this pig-meat! This arises from want of the early habit of attention to such matters. I, who used to get hog-weed for pigs and for rabbits when a little chap, have never forgotten that the wild parsnip is good for pigs and rabbits.

When the doe has young ones, feed her most abundantly with all sorts of greens and herbage, and with carrots and the other things mentioned before, besides giving her a few oats once a day. That is the way to have fine healthy young ones, which, if they come from the mother in good case, will very seldomê die. But do not think that because she is a small animal, a little feeding is sufficient. Rabbits eat a great deal more than cows or sheep in proportion to their bulk.

Of all animals rabbits are those that boys are most fond of. They are extremely pretty, nimble in their movements, engaging in their attitudes, and always completely under immediate control. The produce has not long to be waited for. In short, they keep an interest constantly alive in a little chap's mind; and they really cost nothing; for as to the oats, where is the boy that cannot, in harvest-time, pick up enough along the lanes to serve his rabbits for a year? The care is all; and the habit of taking care of things is, of itself, a most valuable possession.

To those gentlemen who keep rabbits for the use of their family (and a very useful and convenient article they are) I would observe, that when they find their rabbits die, they must depend on it that ninety-nine times out of a hundred starvation is the malady. And particularly short feeding of the doe, while and before she has young ones; that is to say, short feeding of her at all times; for, if she be poor, the young ones will be good for nothing. She will live being poor, but she will not and cannot breed up fine young ones.

GOATS

In some places where a cow cannot be kept, a goat may. A correspondent points out to me, that a Dorset ewe or two might be kept on a common near a cottage to give milk; and certainly this might be done very well; but I should prefer a goat, which is hardier and much more domestic. When I was in the army, in New Brunswick, where, be it observed, the snow lies on the ground seven months in the year, there were many goats that belonged to the regiment, and that went about with it on shipboard and everywhere else. Some of them had gone through nearly the whole of the American War. We never fed them. In summer they picked about wherever they could find grass; and in winter they lived upon cabbage-leaves, turnip-peelings, potato-peelings, and other things flung out of the soldiers' rooms and huts. One of these goats belonged to me, and, on an average throughout the year, she gave me more than three half-pints of milk a day. I used to have the kid killed when a few days old; and, for some time, the goat would give, nearly or quite, two quarts of milk a day. She was seldom dry more than three weeks in the year.

There is one great inconvenience belonging to goats; that is, they bark all young trees that they come near; so that, if they get into a garden, they destroy everything. But there are seldom trees on commons, except such as are too large to be injured by goats; and I can see no reason against keeping a goat where a cow cannot be kept. Nothing is so hardy; nothing is so little nice as to its food. Goats will pick peelings out of the kennel and eat them. They will eat mouldy bread or biscuit; fusty hay, and almost rotten straw; furze-bushes, heath thistles, and, indeed, what will they not eat, when they will make a hearty meal on paper, brown or white, printed on or not printed on, and give milk the while! They will lie in any dog-hole. They do very well clogged, or stumped out. And then, they are very healthy things into the bargain, however closely they may be confined. When sea voyages are so stormy as to kill geese,

ducks, fowls and almost pigs, the goats are well and lively; and when a dog of no kind can keep the deck for a minute, a goat will skip about upon it as bold as brass.

Goats do not ramble from home. They come in regularly in the evening, and if called, they come, like dogs. Now, though ewes, when taken great care of, will be very gentle, and though their milk may be rather more delicate than that of a goat, the ewes must be fed with nice and clean food, and they will not do much in the milk-giving way upon a common; and, as to feeding them provision must be made pretty nearly as for a cow. They will not endure confinement like goats; and they are subject to numerous ailments that goats know nothing of. Then the ewes are done by the time they are about six years old; for they then lose their teeth; whereas the goat will continue to breed and to give milk in abundance for a great many years. The sheep is frightened at everything, and especially at the least sound of a dog. A goat, on the contrary, will face a dog, and, if he be not a big and courageous one, beat him off.

I have often wondered how it happened that none of our labourers kept goats; and I really should be glad to see the thing tried. They are pretty creatures, domestic as a dog, will stand and watch, as a dog does, for a crumb of bread, as you are eating; give you no trouble in the milking; and I cannot help being of opinion, that it might be of great use to introduce them amongst our labourers.

BREWING BEER

In former times, to set about to show to Englishmen that it was good for them to brew beer in their houses, would have been as impertinent as gravely to insist that they ought to endeavour not to lose their breath; for, in those times (only forty years ago), to have a house and not to brew was a rare thing indeed. Mr. Ellman, an old man and a large farmer, in Sussex, has recently given in evidence, before a Committee of the House of Commons, this fact; that forty years ago, there was not a labourer in his parish that did not brew his own beer; and that now there is not one that does it, except by chance the malt be given him The causes of this change have been the lowering of the wages of labour compared with the price of provisions, by the means of the paper-money; the enormous tax upon the barley when made into malt; and the increased tax upon hops. These have quite changed the customs of the English people as to their drink. They still drink beer, but in general it is of the brewing of common brewers and in public-houses, of which the common brewers have become the owners, and have thus, by the aid of paper-money, obtained a monopoly in the supplying of the great body of the people with one of those things which, to the hard-working man, is almost a necessary of life.

These things will be altered. They must be altered. The nation must be sunk into nothingness, or a new system must be adopted; and the nation will not sink into nothingness. The malt now pays a tax of 4s. 6d. a bushel, and the barley costs only 3s. This brings the bushel of malt to 8s. including the maltster's charge for malting. If the tax were taken off the malt, malt would be sold, at the present price of barley, for about 3s. 3d. a bushel; because a bushel of barley makes more than a bushel of malt, and the tax, besides its amount, causes great expenses of various sorts to the maltster. The hops pay a tax of 2d. a pound; and a bushel of malt requires, in general, a pound of hops; if these two taxes were taken off, therefore, the consumption of barley and of hops would be exceedingly increased; for double the present quantity would be demanded, and the land is always ready to send it forth.

It appears impossible that the landlords should much longer submit to these intolerable burdens on their estates. In short, they must get off the malt tax or lose those estates. They must do a great deal more, indeed; but that they must do at any rate. The paper-money is fast losing its destructive power; and things are, with regard to the labourers, coming back to what they were forty years ago, and therefore we may prepare for the making of beer in our own houses, and take leave of the poisonous stuff served out to us by common brewers. We may begin immediately; for, even at present prices, home brewed beer is the cheapest drink that a family can use, except milk, and milk can be applicable only in certain cases.

The drink which has come to supply the place of beer has, in general, been tea. It is notorious that tea has no useful strength in it; that it contains nothing nutritious; that it, besides being good for nothing, has badness in it, because it is well known to produce want of sleep in many cases, and in all cases, to shake and weaken the nerves. It is, in fact, a weaker kind of laudanum, which enlivens for the moment and deadens afterwards. At any rate it communicates no strength to the body; it does not in any degree assist in affording what labour demands. It is, then, of no use. And now, as to its cost, compared with that of beer. I shall make my comparison applicable to a year, or three hundred and sixty-five days. I shall suppose the tea to be only five shillings the pound, the sugar only sevenpence, the milk only twopence a quart. The prices are at the

very lowest. I shall suppose a tea-pot to cost a shilling, six cups
and saucers two shillings and sixpence, and six pewter spoons
eighteen-pence. How to estimate the firing I hardly know, but
certainly there must be in the course of the year two hundred
fires made that would not be made, were it not for tea drinking.
Then comes the great article of all, the time employed in this
tea-making affair. It is impossible to make a fire, boil water,
make the tea, drink it, wash up the things, sweep up the fire-
place and put all to rights again in a less space of time, upon
the average, than two hours. However, let us allow one hour;
and here we have a woman occupied no less than three hundred
and sixty-five hours in the year; or thirty whole days at twelve
hours in the day; that is to say, one month out of the twelve in the
year, besides the waste of the man's time in hanging about waiting
for the tea! Needs there anything more to make us cease to
wonder at seeing labourers' children with dirty linen and holes
in the heels of their stockings? Observe too, that the time thus
spent is, one half of it, the best time of the day. It is the top of
the morning, which, in every calling of life, contains an hour
worth two or three hours of the afternoon. By the time that the
clattering tea-tackle is out of the way, the morning is spoiled,
its prime is gone, and any work that is to be done afterwards
lags heavily along. If the mother have to go out to work the tea
affair must all first be over. She comes into the field, in summer
time, when the sun has gone a third part of his course. She has
the heat of the day to encounter, instead of having her work
done and being ready to return home at an early hour. Yet
early she must go too; for there is the fire again to be made, the
clattering tea-tackle again to come forward; and even in the
longest day she must have candle light, which never ought to
be seen in a cottage (except in case of illness) from March to
September.

Now, then, let us take the bare cost of the use of tea. I suppose a pound of tea to last twenty days, which is not nearly half
an ounce each morning and evening. I allow for each mess half
a pint of milk. And I allow three pounds of the red dirty sugar

to each pound of tea. The account of expenditure would then stand very high; but to these must be added the amount of the tea-tackle, one set of which will, upon an average, be demolished every year. To these outgoings must be added the cost of beer at the public-house; for some the man will have, after all, and the woman too, unless they be upon the point of actual starvation. Two pots a week is as little as will serve in this way; and here is a dead loss of ninepence a week, seeing that two pots of beer, full as strong, and a great deal better, can be brewed at home for three pence. The account of the year's tea drinking will then stand thus:

	£	s	d
18-lb. of tea	4	10	0
54-lb. of sugar	1	11	6
365 pints of milk	1	10	0
Tea-tackle		5	0
200 fires		16	8
30 days' work		15	0
Loss by going to public-house...	1	19	0
	£11	7	2

I have here estimated everything at its very lowest. The entertainment which I have here provided is as poor, as mean, as miserable, as any thing short of starvation can set forth; and yet the wretched thing amounts to a good third part of a good and able labourer's wages! For this money he and his family may drink good and wholesome beer; in a short time, out of the mere savings from this waste, may drink it out of silver cups and tankards. In a labourer's family, wholesome beer, that has a little life in it, is all that is wanted in general. Little children, that do not work, should not have beer. Broth, porridge, or something in that way, is the thing for them. However, I shall suppose, in order to make my comparison as little complicated as possible, that he brews nothing but beer as strong as the

generality of beer to be had at the public-house, and divested
of the poisonous drugs which that beer but too often contains;
and I shall further suppose that he uses in his family two quarts
of this beer every day from the first of October to the last day of
March inclusive; three quarts a day during the months of June
and September; and five quarts a day during the months of
July and August; and if this be not enough, it must be a family
of drunkards. Here are 1,097 quarts or 274 gallons. Now, a
bushel of malt will make eighteen gallons of better beer than
that which is sold at the public-houses. And this is precisely a
gallon for the price of a quart. People should bear in mind, that
the beer bought at the public-house is loaded with a beer tax,
with the tax on the public-house keeper, in the shape of license,
with all the taxes and expenses of the brewer, and with all taxes,
rent, and other expenses of the publican, and with all the profits
of both brewer and publican; so that when a man swallows a
pot of beer at a public-house, he has all these expenses to help
to defray, besides the mere tax on the malt and on the hops.

Well, then, to brew this ample supply of good beer for a
labourer's family, these 274 gallons, requires fifteen bushels
of malt and (for let us do the thing well) fifteen pounds of hops.
The malt is now eight shillings a bushel, and very good hops
may be bought for less than a shilling a pound. The grains and
yeast will amply pay for the labour and fuel employed in the
brewing; seeing that there will be pigs to eat the grains, and
bread to be baked with the yeast. The account will then stand
thus:

	£	s	d
15 bushels of malt	6	0	0
15 pounds of hops		15	0
Wear of utensils		10	0
	£7	5	0

Here, then, is the sum of four pounds two shillings and two

pence saved every year. The utensils for brewing are, a brass kettle, a mashing tub, coolers (for which washing tubs may serve), a half hogs-head, with one end taken out, for a tun tub, about four nine-gallon casks, and a couple of eighteen-gallon casks. This is an ample supply of utensils, each of which will last, with proper care, a good long lifetime or two, and the whole of which, even if purchased new from the shop, will only exceed by a few shillings if they exceed at all, the amount of the saving, arising *the very first year*, from quitting the troublesome and pernicious practice of drinking tea. The saving of each succeeding year would, if you chose it, purchase a silver mug to hold half a pint at least. However, the saving would naturally be applied to purposes more conducive to the well-being and happiness of a family.

It is not, however, the mere saving to which I look. This is, indeed, a matter of great importance, whether we look at the amount itself, or at the ultimate consequence of a judicious application of it; for four pounds make a great hole in a man's wages for the year; and when we consider all the advantages that would arise to a family of children from having these four pounds, now so miserably wasted, laid out upon their backs, in the shape of a decent dress, it is impossible to look at this waste without feelings of sorrow, not wholly unmixed with those of a harsher description.

But I look upon the thing in a still more serious light. I view the tea drinking as a destroyer of health, an enfeebler of the frame, an engenderer of effeminacy and laziness, a debaucher of youth and a maker of misery for old age. In the fifteen bushels of malt there are 570 pounds weight of sweet; that is to say, of nutritious matter, unmixed with any thing injurious to health. In the 730 tea messes of the year there are 54 pounds of sweet in the sugar and about 30 pounds of matter equal to sugar in the milk. Here are eighty-four pounds instead of five hundred and seventy, and even the good effect of these eighty-four pounds is more than over-balanced by the corrosive, gnawing, and poisonous powers of the tea.

It is impossible for any one to deny the truth of this statement. Put it to the test with a lean hog; give him the fifteen bushels of malt, and he will repay you in ten score of bacon or thereabouts. But give him the 730 tea messes, or rather begin to give them to him, and give him nothing else, and he is dead with hunger, and bequeaths you his skeleton, at the end of about seven days. It is impossible to doubt in such a case. The tea drinking has gone a great deal in bringing this nation into the state of misery in which it now is; and the tea drinking, which is carried on by 'dribs' and 'drabs' by pence and farthings going out at a time; this miserable practice has been gradually introduced by the growing weight of the taxes on malt and on hops and by the everlasting penury amongst the labourers, occasioned by the paper-money.

We see better prospects, however, and therefore let us now rouse ourselves, and shake from us the degrading curse, the effects of which have been much more extensive and infinitely more mischievous than men in general seem to imagine.

It must be evident to every one, that the practice of tea drinking must render the frame feeble and unfit to encounter hard labour or severe weather, while, as I have shown, it deducts from the means of replenishing the belly and covering the back. Hence succeeds a softness, an effeminacy, a seeking for the fire-side, a lurking in the bed, and, in short, all the characteristics of idleness, for which, in this case, real want of strength furnishes an apology. The tea drinking fills the public-house, makes the frequenting of it habitual, corrupts boys as soon as they are able to move from home, and does little less for the girls, to whom the gossip of the tea-table is no bad preparatory school for the brothel. At the very least, it teaches them idleness. The everlasting dawdling about with the slops of the tea-tackle gives them a relish for nothing that requires strength and activity. When they go from home, they know how to do nothing that is useful. To brew, to bake, to make butter, to milk, to rear poultry; to do any earthly thing of use they are wholly unqualified. To shut poor young creatures up in manufactories is bad enough; but there, at any rate, they do something that is useful; whereas the girl that has been brought up merely to boil the tea-kettle, and to assist in the gossip inseparable from the practice, is a mere consumer of food, a pest to her employer, and a curse to her husband, if any man be so unfortunate as to fix his affections upon her.

But is it in the power of any man, any good labourer who has attained the age of fifty, to look back upon the last thirty years of his life, without cursing the day in which tea was introduced into England? Where is there such a man, who cannot trace to this cause a very considerable part of all the mortifications and sufferings of his life? When was he ever too late at his labour; when did he ever meet with a frown, with a turning off, and pauperism on that account, without being able to trace it to the tea-kettle? When reproached with lagging in the morning, the poor wretch tells you that he will make up for it by working during his breakfast time! I have heard this a hundred and a hundred times over. He was up time enough; but the tea-kettle kept him lolling and lounging at home; and now instead of sitting down to a breakfast upon bread, bacon and beer, which is to carry him on to the hour of dinner, he has to force his limbs along under the sweat of feebleness, and at dinner-time to swallow his dry bread, or slake his half-feverish thirst at the pump or the brook. To the wretched tea-kettle he has to return at night, with legs hardly sufficient to maintain him; and thus he makes his miserable progress towards that death which he finds ten or fifteen years sooner than he would have found it had he made his wife brew beer instead of making tea. If he now and then gladdens his heart with the drugs of the public-house, some quarrel, some accident, some illness is the probable consequence; to the affray abroad succeeds the affray at home; the mischievous example reaches the children, corrupts them or scatters them, and misery for life is the consequence.

RURAL SPORTS

RURAL SPORTS

THE GAME

The great business of life, in the country, appertains, in some way or other, to the *game*, and especially at this time of the year*. If it were not for the game, a country life would be like an *everlasting honey-moon*, which would, in about half a century, put an end to the human race. In *towns*, or large villages, people make a shift to find the means of rubbing the rust off from each other by a vast variety of sources of contest. A couple of wives meeting in the street, and giving each other a wry look, or a look not quite civil enough, will, if the parties be hard pushed for a ground of contention, do pretty well. But in the country, there is, alas! no such resource. Here are no walls for people to take of each other. Here they are so placed as to prevent the possibility of such lucky local contact. Here is more than room of every sort, elbow, leg, horse, or carriage, for them all. Even *at Church* (most of the people being in the meeting-houses) the pews are surprisingly too large. Here, therefore, where all circumstances seem calculated to cause never-ceasing concord with its accompanying dullness, there would be no relief at all, were it not for the *game*. This, happily, supplies the place of all other sources of alternate dispute and reconciliation; it keeps all in life and motion, from the lord down to the hedger. When I see two men, whether in a market-room, by the way-side, in a parlour, in a church yard, or even in the church itself, engaged in manifestly deep and

* October (1825)

most momentous discourse, I will, if it be any time between September and February, bet *ten to one*, that it is, in some way or other, about *the game*. The wives and daughters hear so much of it, that they inevitably get engaged in the disputes; and thus all are kept in a state of vivid animation. I should like very much to be able to take a spot, a circle of 12 miles in diameter, and take an exact account of all the *time* spent by each individual, above the age of *ten* (that is the age they begin at), in *talking*, during the game season of one year, about the *game* and about *sporting exploits*. I verily believe that it would amount, upon an average, to *six times* as much as *all the other talk put together*; and, as to the *anger*, the *satisfaction*, the *scolding*, the *commendation*, the *chagrin*, the *exultation*, the *envy*, the *emulation*, where are there any of these in the country, unconnected with *the game*?

There is, however, an important distinction to be made between *hunters* (including coursers) and *shooters*. The latter are, as far as relates to their exploits, a disagreeable class, compared with the former; and the reason of this is, their doings are almost wholly *their own*; while, in the case of the others, the achievements are the property of *the dogs*. Nobody likes to hear another talk *much* in praise of his own acts, unless those acts have a manifest tendency to produce some good to the hearer; and shooters do talk *much* of their own exploits, and those exploits rather tend to *humiliate* the hearer. Then, a *great shooter* will, nine times out of ten, go so far as almost to *lie a little*; and, though people do not tell him of it, they do not like him the better for it; and he but too frequently discovers that they do not believe him: whereas, hunters are mere followers of the dogs, as mere *spectators*; their praises, if any are called for, are bestowed on the greyhounds, the hounds, the fox, the hare, or the horses. There is a little rivalship in the riding, or in the behaviour of the horses; but this has so little to do with the *personal merit* of the sportsmen, that it never produces a want of good fellowship in the evening of the day. A shooter who has been *missing* all day, must have an uncommon

share of good sense, not to feel mortified while the slaughterers
are relating the adventures of that day; and this is what cannot
exist in the case of the hunters. Bring me into a room, with a
dozen men in it, who have been sporting all day; or, rather let
me be in an adjoining room, where I can hear the sound of
their voices, without being able to distinguish the words, and I
will bet ten to one that I tell whether they be hunters or
shooters.

I was once acquainted with a *famous shooter* whose name was
William Ewing. He was a barrister of Philadelphia, but became
far more renowned by his gun than by his law cases. We spent
scores of days together a-shooting, and were extremely well
matched, I having excellent dogs and caring little about my
reputation as a shot, his dogs being good for nothing, and he
caring more about his reputation as a shot than as a lawyer.
The fact which I am going to relate respecting this gentleman,
ought to be a warning to young men, how they become
enamoured of this species of vanity. We had gone about ten
miles from our home, to shoot where partridges were said to be
very plentiful. We found them so. In the course of a November
day, he had, just before dark, shot, and sent to the farm-house,
or kept in his bag, *ninety-nine* partridges. He made some few
double shots, and he might have a *miss* or two, for he sometimes
shot when out of my sight, on account of the woods. However,
he said that he killed at every shot; and, as he had counted the
birds, when we went to dinner at the farm-house and when he
cleaned his gun, he, just before sun-set, knew that he had
killed *ninety-nine* partridges, every one upon the wing, and a
great part of them in woods very thickly set with largish trees.
It was a grand achievement; but, unfortunately, he wanted to
make it *a hundred*. The sun w*as setting*, and, in that country,
darkness comes almost at once; it is more like the going out of
a candle than that of a fire; and I wanted to be off, as we had
a very bad road to go, and as he, being under strict petticoat
government, to which he most loyally and dutifully submitted,
was compelled to get home that night, taking me with him, the

vehicle (horse and gig) being mine. I, therefore, pressed him to
come away, and moved on myself towards the house (that of
old John Brown, in Bucks county, grandfather of that General
Brown, who gave some of our whiskered heroes such a rough
handling last war, which was waged for the purpose of 'deposing
James Madison'), at which house I would have stayed all
night, but from which I was compelled to go by that watchful
government, under which he had the good fortune to live.
Therefore I was in haste to be off. No: he would kill the
hundredth bird! In vain did I talk of the bad road and its many
dangers for want of moon. The poor partridges, which we had
scattered about, were *calling* all around us; and, just at this
moment, up got one under his feet, in a field in which the
wheat was three or four inches high. He shot and *missed*. 'That's
it,' said he, running as if to *pick up* the bird. 'What!' said I,
'you don't think you *killed*, do you? Why there is the bird now,
not only alive, but *calling*, in that wood'; which was at about a
hundred yards distance. He, in that *form of words* usually employed in such cases, asserted that he shot the bird and saw it
fall; and I, in much about the same form of words, asserted,
that he had *missed*, and that I, with my own eyes, saw the bird
fly into the wood. This was too much! To *miss* once out of a
hundred times! To lose such a chance of immortality! He was
a good-humoured man; I liked him very much; and I could not
help feeling for him, when he said, 'Well, *Sir*, I killed the bird;
and if you choose to go away and take your dog away, so as
to prevent me from *finding* it, you must do it; the dog is *yours*,
to be sure.' 'The *dog*,' said I, in a very mild tone, 'why, Ewing,
there is the spot; and could we not see it, upon this smooth green
surface, if it were there?' However, he began to *look about*; and
I called the dog, and affected to *join him in the search*. Pity for
his weakness got the better of my dread of the bad road. After
walking backward and forward many times upon about twenty
yards square with our eyes to the ground, looking for what both
of us knew was not there, I had *passed him* (he going one way and
I the other), and I happened to be turning round just after I

had passed him, when I saw him, putting his hand behind him, *take a partridge out of his bag and let it fall upon the ground*! I felt no temptation to detect him, but turned away my head, and kept looking about. Presently he, having returned to the spot where the bird was, called out to me, in a most triumphant tone; '*Here! here!* Come here!' I went up to him, and he, pointing with his finger down to the bird, and looking hard in my face at the same time, said, 'There, Cobbett; I hope that will be a *warning* to you never to be obstinate again'! 'Well,' said I, 'come along': and away we went as merry as larks. When we got to Brown's, he told them the story, triumphed over me most clamorously; and, though he often repeated the story to my face, I never had the heart to let him know, that I knew of the imposition, which puerile vanity had induced so sensible and honourable a man to be mean enough to practise. A *professed shot* is, almost always, a very disagreeable brother sportsman. He must, in the first place, have a head rather of the emptiest to *pride himself* upon so poor a talent. Then he is always out of temper, if the game fail, or if he miss it. He never participates in that great delight which all sensible men enjoy at beholding the beautiful action, the docility, the zeal, the wonderful sagacity, of the pointer and the setter. He is always thinking about *himself*; always anxious to surpass his companions. I remember that, once, Ewing and I had lost our dog. We were in a wood, and the dog had gone out, and found a covey in a wheat stubble joining the wood. We had been whistling and calling him for, perhaps, half an hour, or more. When we came out of the wood we saw him pointing, with one foot up; and, soon after, he, keeping his foot and body unmoved, gently turned round his head towards the spot where he heard us, as if to bid us come on, and, when he saw that we saw him, turned his head back again. I was so delighted, that I stopped to look with admiration. Ewing, astonished at my want of alacrity, pushed on, shot one of the partridges, and thought no more about the conduct of the dog than if the sagacious creature had had nothing at all to do with the matter. When I left

America, in 1800, I gave this dog to Lord Henry Stuart, who was, when he came home, a year or two afterwards, about to bring him to astonish the sportsmen even in England; but, those of Pennsylvania were resolved not to part with him, and, therefore they *stole* him the night before his Lordship came away. Lord Henry had plenty of pointers after his return, and he *saw* hundreds; but always declared, that he never saw any thing approaching in excellence this American dog. For the information of sportsmen I ought to say, that this was a small-headed and sharp-nosed pointer, hair as fine as that of a greyhound, little and short ears, very light in the body, very long legged, and swift as a good lurcher. I had him a puppy, and he never had any *breaking*, but he pointed staunchly at once; and I am of opinion, that this sort is, in all respects, better than the heavy breed. Mr Thornton, (I beg his pardon, I believe he is now a *Knight* of some sort) who was, and perhaps still is, our *Envoy in Portugal*, and who, at the time here referred to, was a sort of *partner* with Lord Henry in this famous dog; and gratitude (to the memory of *the dog* I mean,) will, I am sure, or, at least, I hope so, make him bear witness to the truth of my character of him; and, if one could hear an Ambassador *speak out*, I think that Mr Thornton would acknowledge, that his calling has brought him in pretty close contact with many a man who was possessed of most tremendous political power, without possessing half the sagacity, half the understanding, of this dog, and without being a thousandth part so faithful to his trust. I am quite satisfied, that there are as many *sorts* of men as there are of dogs. Swift was a man, and so is Walter the base. But, is the *sort* the same? It cannot be *education* alone that makes the amazing difference that we see. Besides, we see men of the very same rank and riches and education, differing as widely as the pointer does from the pug. The name, *man*, is common to all the sorts, and hence arises very great mischief. What confusion must there be in rural affairs, if there were no names whereby to distinguish hounds, greyhounds, pointers, spaniels, terriers, and sheep dogs, from each other! And, what pretty

work, if, without regard to the *sorts* of dogs, men were to attempt
to *employ them*! Yet, this is done in the case of *men*! A man is
always *a man*; and, without the least regard as to the *sort*, they
are promiscuously placed in all kinds of situations. Now, if
Mr Brougham, Doctors Birkbeck, MacCulloch and Black, and
that profound personage, Lord John Russell, will, in their forth-
coming 'London University,' teach us how to divide men *into
sorts*, instead of teaching us to *augment* the CAPITAL *of the nation
by making paper-money*, they will render us a real service. That
will be *feelosofy* worth attending to. What would be said of the
'Squire who should take a fox-hound out to find partridges for
him to shoot at? Yet, would this be *more* absurd than to set a
man to law-making, who was manifestly formed for the express
purpose of sweeping the streets or digging out sewers?

IN DEFENCE OF BLOOD SPORTS

There are persons who question *the right* of man to pursue and
destroy the wild animals, which are called *game*. Such persons
however claim the right of killing *foxes* and *hawks*; yet these
have as much right to live and to follow their food as *pheasants*
and *partridges* have. This, therefore, in such persons, is *nonsense*.

Others, in their mitigated hostility to the sports of the field,
say that it is *wanton* cruelty to shoot or hunt; and that we kill
animals from the farmyard only because their flesh is *necessary
to our own existence*. PROVE THAT. No: you cannot. If you
could it is but the *'tyrant's* plea;' but you cannot; for we know
that men can and do live without animal food and if their
labour be not of an exhausting kind, live well too, and longer
than those who eat it. It comes to this then, that we kill hogs
and oxen because we *choose* to kill them; and we kill game for
precisely the same reason.

A third class of objectors, seeing the weak position of the two
former, and still resolved to eat flesh, take their stand upon this
ground: that sportsmen send some game off *wounded* and leave
them in a *state of suffering*. These gentlemen forget the operations
performed on calves, pigs, lambs and sometimes on poultry.

Sir Isaac Coffin prides himself upon teaching the English ladies
how to make *turkey-capons*. Only think of the separation of calves,
pigs and lambs at an early age from their mothers! Go you
sentimental eaters of veal, sucking pig and lamb and hear the
mournful lowings, whinings and bleatings; observe the anxious
listen, the wistful look and the dropping tear of the disconsolate
dams; and then while you have the carcasses of their young
ones under your teeth, cry out as soon as you can empty your
mouths a little, against the *cruelty* of hunting and shooting. Get
up from dinner (but take care to stuff well first) and go and
drown the puppies of the bitch and the kittens of the cat, lest
they should share a little in what their mothers have guarded
with so much fidelity; and as good stuffing may tend to make
you restless in the night, order the geese to be picked alive that
however your consciences may feel, your bed, at least, may be
easy and soft. Witness all this with your own eyes, and then
go weeping to bed at the possibility of a hare having been
terribly frightened without being killed, or of a bird having
been left in a thicket with a shot in its body or a fracture in its
wing. But before you go up stairs, give your servants orders to
be early at market for fish, fresh out of the water; that they may
be *scaled* or *skinned alive!* A truce with you then, sentimental
eaters of flesh; and here I propose the terms of a lasting com-
promise with you. We must on each side yield something; we
sportsmen will content ourselves with merely *seeing the hares
skip and the birds fly*; and you shall be content with the flesh
and fish that come from cases of *natural death* of which I am
sure your compassionate disposition will not refuse us a trifling
allowance.

Nor have even the *Pythagoreans* a much better battery against
us. Sir Richard Phillips who once rang a peal in my days against
shooting and hunting, does indeed eat neither *flesh*, *fish* nor
fowl. His abstinence surpasses that of a Carmelite, while his
bulk would not disgrace a Benedictine Monk or a Protestant
Dean. But he forgets that his *shoes* and *breeches* and *gloves* are
made with the skins of animals; he forgets that he *writes* (and

very eloquently too) with what has been cruelly taken from a fowl; and that in order to cover the *books* which he has had made and sold, hundreds of flocks and scores of droves must have perished; nay, that to get him his *beaver-hat* a beaver must have been *hunted* and killed and in the doing of which many beavers have been *wounded* and left to pine away the rest of their lives; and perhaps many little orphan beavers, left to lament the murder of their parents. Bewley was the only real and sincere Pythagorean of modern times that I ever heard of. He protested not only against eating the flesh of animals but also against robbing their backs; and therefore his dress consisted wholly of flax. But even he, like Sir Richard Phillips, ate milk, butter, cheese and eggs though this was cruelly robbing the hens, cows and calves; and indeed causing the murder of calves. In addition poor little Ben forgot the materials of *book-binding*; and it was well he did for else his Bible would have gone into the fire!

Taking it for granted then that sportsmen are as good as other folks on the score of *humanity*, the sports of the field, like everything else done in the fields, tend to produce or preserve *health*. I prefer them to all other pastimes, because they produce *early rising*; because they have no tendency to lead young men into vicious habits. It is where men congregate that the vices haunt. A hunter or a shooter may also be a gambler or a drinker; but he is less likely to be fond of the two latter if he be fond of the former. Boys will take to something in the way of pastime; and it is better that they take to that which is innocent, healthy and manly than that which is vicious, unhealthy and effeminate. Besides, the scenes of rural sports are necessarily at *a distance from cities and towns*. This is another great consideration; for though great talents are wanted to be *employed* in the *hives of men*, they are very rarely *acquired* in these hives; the surrounding objects are too numerous, too near the eye, too frequently under it and too artificial.

For these reasons I have always encouraged my sons to pursue these sports. They have, until the age of 14 or 15, spent

their time by day chiefly amongst horses and dogs and in the
fields and farm-yard; and their candlelight has been spent
chiefly in reading books about hunting and shooting and about
dogs and horses. I have supplied them plentifully with *books*
and *prints* relating to these matters. They have *drawn* horses,
dogs and game themselves. These things, in which they took so
deep an interest, not only engaged their attention and wholly
kept them from all taste for and even all knowledge of cards and
other senseless amusements, but they led them *to read and write
of their own accord; and never in my life have I set them a copy in
writing nor attempted to teach them a word of reading.* They have
learnt to read by looking into books about dogs and game;
and they have learnt to write by imitating my writing, and by
writing endless letters to me when I have been from home about
their dogs and other rural concerns. While the Borough-tyrants
had me in Newgate for two years with a thousand men in the
heart of England, under a guard of Hanoverian sabers, I
received *volumes of letters* from my children; and I have them
now, from the *scrawl* of *three years* to the neat and beautiful hand
of thirteen. I never told them of any *errors* in their letters. All
was well. The best evidence of the utility of their writing and
the strongest encouragement to write again was *a very clear
answer from me* in a very precise hand and upon very nice paper
which they never failed promptly to receive. They have all
writen to me *before they could form a single letter.* A little bit of
paper with some ink-marks on it folded up by themselves and
a wafer stuck to it used to be sent to me and it was sure to bring
the writer a very, very kind answer. Thus have they gone on.
So far from being a *trouble* to me they have been all pleasure
and advantage. For many years they have been so many
secretaries. I have dictated scores of registers to them, which
have *gone to the press without my even looking at them.* I dictated
registers to them at the age of *thirteen* and even of *twelve.* They
have, as to *trustworthiness*, been grown persons at eleven or
twelve. I could leave my house and affairs, the paying of men
or the going from home on business, to them at an age when

boys in England, in general, want servants to watch them to see that they do not kill chickens, torment kittens or set the buildings on fire.

Here is a good deal of *boasting*; but it will not be denied that I have *done a great deal* in a short public life and I see no harm in telling my readers of any of the means that I have employed; especially as I know of few greater misfortunes than that of breeding up things to be *schoolboys all their lives*. It is not that I have so many wonders of the world; it is that I have pursued a rational plan of education and one that any man may pursue, if he will, with similar effects. I remembered too that I myself had had a sportsman-education. I ran after the hare-hounds at the age of *nine or ten*. I have many and many a day left the rooks to dig up the wheat and peas. while I followed the hounds and have returned home at dark-night with my legs full of thorns and my belly empty to go supperless to bed and to congratulate myself if I escaped a flogging. I was sure of these consequences; but that had not the smallest effect in restraining me. All the lectures, all the threats vanished from my mind in a moment upon hearing the first cry of the hounds, at which my heart used to be ready to bound out of my body. I *remembered* all this. I traced to this taste my contempt for card playing and for all childish and effeminate amusements. And therefore I resolved to leave the same course freely open to my sons. This is *my plan of education*; others may follow what plan they please.

Letter to William Windham, 6 October 1805:

> . . . I have not the confidence to hope that you will come all the way to Botley to see the sport, of which the enclosed paper speaks; but, if you should, I think you will be highly gratified. We expect 5000 people. Certainly 4000 more than there are houses to hold. But *my* house (for such I shall soon be able to make it) will contain a room for you, with everything about it tolerably decent and comfortable. At any rate, Southampton is very near. This is an exercise that requires great strength and very great fortitude. The

SINGLE STICK
SINGLE-STICK PLAYING
AT BOTLEY, NEAR SOUTHAMPTON
On Friday, October the 11th, 1805
(Being old Michaelmas-day)
Will be played in the village of
BOTLEY
A Grand Match at Single-Stick.
The prizes will be as follows:
1st Prize. —Fifteen Guineas and a Gold-laced Hat.
2nd Prize.—Six Guineas and a Silver-laced Hat.
3rd Prize. —Four Guineas.
4th Prize. —Two Guineas.

The Terms, as to playing, and the Ties, &c., will be
announced upon the spot. Those who have played for, and
lost, the First Prize, will be allowed to play for the Second:
Those who have lost the Second, will be allowed to play
for the Third; and those who have lost the Third will be
allowed to play for the Fourth. The Playing will begin at
Eleven o'clock in the Morning, and, if possible, all the
Prizes are to be played for on the same Day. For any further
Information that may be required, application may be
made, either in Person, or by Letter, to Mr Richard Smith,
of Botley. Gentlemen coming from a distance will find
Excellent Accommodations, of every kind, at and in the
neighbourhood of Botley, which is situated at only about
Five Miles from Southampton, and at less than Four
Miles from Bishop's Waltham; the distance from London
through Farnham, Alton, and Bishop's Waltham, is a short
day's journey, being barely Sixty-Eight Miles.
BOTLEY, *September* 23, 1805.

J. Brettell, Printer, Marshall Street, Golden Square, London players use a stick three-quarters of an inch in diameter, $2\frac{1}{2}$ feet long, and having a basket hilt to defend the hand. They are stripped to the shirt; and the object is to break the antagonist's head in such a way that the *blood may run an inch*. The blows that they interchange, in order to throw one another off their guard, are such as require the utmost degree of patient endurance. The arms, shoulders, and ribs are beaten black and blue, and the contest between two men frequently lasts for more than an hour. Last Whitsuntide there was a match at Bishop's Waltham, where one of the players, feeling that he had a tooth knocked out, and knowing that if he opened his mouth the blood would be perceived, *swallowed both blood and tooth*, and continued the combat (with two others driven from their places in his gum) till he obtained the victory. And this only for *a prize of a guinea!* We expect players from All the Western Counties on this side of Cornwall. Our advertisements have roused all this part of the country completely. The prizes and other expenses are to be defrayed by about 18 farmers and millers and myself. The whole will amount to about two guineas each. What a miserable sum wherewith to produce such an effect! The *justices* here are rather a harmless caste; and, at any rate, they love their comfort and the goodwill of their neighbours too much to attempt to interfere with us.

Letter to William Windham, 15 October 1805

. . . Before this reaches you, you will probably have seen, in the *Morning Chronicle*, a pretty full account of our sports; and, therefore, I will only say here, that there has not been, in the memory of any one here, such a match at Single-stick in Hampshire. You will see that we had seven players up from the West. They carried off three of our prizes, but they have also taken away some most memorable proofs of Hampshire hardihood. You would have been delighted to witness the peaceableness and even the silence of the people. The village was full; and, which was matter of no small satisfaction to me, several gentlemen of the neighbourhood, amongst whom was Sir Joseph Sidney Yorke (Mr Yorke's brother), after having been spectators of the sport, begged to be admitted as patrons of it. I took the most effectual way. *First*, I published my advertisement, resolved that the playing should take place, and then I sent round to offer people an opportunity of partaking with me in the patronage of it. The overplus money thrown in upon us was very considerable in amount, and the farmers agreed with me in giving it to the players of the West who won no prizes, and who did, indeed, stand in need of something to comfort them.

HARE COURSING

22 October 1822

Before you get to Salisbury, you cross the valley that brings down a little river from *Amesbury*. It is a very beautiful valley. There is a chain of farm-houses and little churches all the way up it. The farms consist of the land on the flats on each side of the river, running out to a greater or less extent, at different places, towards the hills and downs. Not far above *Amesbury* is a little village called *Netherhaven*, where I once saw an *acre of hares*. We were coursing at *Everly*, a few miles off; and, one of the party happening to say, that he had seen *an acre of hares* at Mr *Hicks Beech's* at Netherhaven, we, who wanted to see the same, or to detect our informant, sent a messenger to beg a day's coursing, which being granted, we went over the next day. Mr Beech received us very politely. He took us into a wheat stubble close by his paddock; his son took a gallop round, cracking his whip at the same time; the hares (which were very thickly in sight before) started all over the field, ran into a *flock* like sheep; and we all agreed, that the flock did cover *an acre of ground*. Mr Beech had an

old greyhound, that I saw lying down in the shrubbery close
by the house, while several hares were sitting and skipping
about, with just as much confidence as cats sit by a dog in
a kitchen or a parlour. Was this *instinct* in either dog or
hares? Then, mind, this same greyhound went amongst
the rest to *course* with us out upon the distant hills and lands;
and then he ran as eagerly as the rest, and killed the hares
with as little remorse. Philosophers will talk a long while
before they will make men believe, that this was *instinct
alone*. I believe that this dog had much more reason than
one half of the Cossacks have; and I am sure he had a
great deal more than many a Negro that I have seen.

27 August 1826

Everley is but about three miles from *Ludgarshall*, so that
we got here in the afternoon of Friday; and, in the evening a very heavy storm came and drove away all flies, and
made the air delightful. This is real *Down*-country. Here
you see miles and miles square without a tree, or hedge,
or bush. It is country of *greensward*. This is the most famous
place in all England for *coursing*. I was here, at this very
inn, with a party *eighteen years ago*; and, the landlord, who
is still the same, recognized me as soon as he saw me. There
were *forty brace of greyhounds* taken out into the fields on
one of the days, and every brace had one course, and some
of them two. The ground is the finest in the world; from
two to three miles for the hare to run to cover, and not a
stone nor a bush nor a hillock. It was here proved to me,
that the hare is, by far, the swiftest of all English animals;
for I saw three hares, in one day *run away* from the dogs.
To give dog and hare a fair trial, there should be but *one*
dog. Then, if that dog got so close as to compel the hare *to
turn*, that would be a proof that the dog ran fastest. When
the dog, or dogs, never get near enough to the hare to induce her to *turn*, she is said, and very justly, to '*run away*'
from them; and, as I saw three hares do this in one day, I

conclude, that the hare is the swiftest animal of the two.

HARE HUNTING
16 November 1821
Ross-on-Wye

A whole day most delightfully passed at hare hunting with a pretty pack of hounds kept here by Messrs Palmer. They put me upon a horse that seemed to have been made on purpose for me, strong, tall, gentle and bold; and that carried me either over or through everything. I, who am just the weight of a four-bushel sack of good wheat, actually sat on his back from daylight in the morning to dusk (about nine hours), without once setting my foot on the ground. Our ground was at Orcop, a place about four miles distance from this place. We found a hare in a few minutes after throwing off; and, in the course of the day, we had to find four, and were never more than ten minutes in finding. A steep and naked ridge, lying between two fat valleys, having a mixture of pretty large fields and small woods, formed our ground. The hares crossed the ridges forward and backward, and gave us numerous views and very fine sport – I never rode on such steep ground before; and, really, in going up and down some of the craggy places where the rain had washed the earth from the rocks, I did think, once or twice of my neck, and how Sidmouth* would like to see me. – As to the *cruelty*, as some pretend of the sport, that point I have I think, settled, in one of the chapters of my *Year's Residence in America*. As to the expense, a pack, even a full pack of harriers, like this, costs less than two bottles of wine a day with their inseparable concomitants. And, as to the *time* thus spent, hunting is inseparable from *early rising*; and, with habits of early rising, who ever wanted time for any business?

*Lord Sidmouth (1757–1844), Home Secretary in 1817 whose 'gaging acts' caused Cobbett to leave for America

TWO SWEETCORN RECIPES

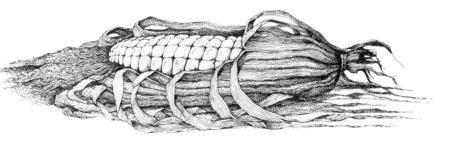

TWO SWEETCORN RECIPES

MUSH

This is not a word to squall out over a piano-forte; but, it is a very good word, and a real *English* word, though Johnson has left it, as he has many other good words, out of his Dictionary. It means this; you put some water or milk into a pot, and bring it to the boil, you then let the flour or meal out of one hand gently into the milk or water, keeping stirring with the other, until you have got it into a pretty stiff state, after which, you let it stand ten minutes, or a quarter of an hour, or less, or even only one minute, and then take it out, and put it into a dish or bowl. This sort of half pudding half porridge you eat either hot or cold, with a little salt or without it. It is frequently eaten unaccompanied with any liquid matter; but the general way is to have a porringer of milk and, taking off a lump of the mush at the time, and putting it into the milk, you take up a spoonful at a time, having a little milk along with it; and this is called *mush* and *milk*. But, here is a most excellent pudding, even if there be no milk to eat with it, that is to say, if it be made originally with milk; and, if there be no milk to be had, as must be generally the case at the houses of the labourers, here is a very good *substitute* for bread, whether you take it cold or hot. It is not, like the miserable potatoe, a thing that turns immediately to water; nor is it like pudding made of *flour* and *water,* which is hard, closely clung together, heavy upon the stomach, indigestible, and of course unwholesome, whence comes the old saying, "Cold pudding to *settle* your love;" that is to say, to cool a fellow exceedingly, if not to extinguish the source of his passions altogether. The *mush* so far from being hard and lumpy when cold, is quite light, very much puffed up; and this is the very thing, made of water, and not of milk, which physicians, recommend to all persons, who from over eating, over drinking, or any other cause, have feeble stomachs. The corn-meal and flour is wholesome, more so than wheat flour, in all its states or manners of cooking; but, this is the manner the most in vogue throughout the United States of America; and, if a poor man's family had plenty of this, even without the milk, he never ought to regret the absence of bread. One great convenience belonging to the mush is, that you may eat it cold, and it most frequently is eaten cold. It may be carried by the workman to the field in a little tin or pewter thing. It is, in fact, moist bread; habit soon makes it as pleasant and even pleasanter than bread. You cannot make mush of wheat-meal nor of oat-meal. It is better to make it of meal that has nothing but the very coarse bran taken out of it, than it is to make it of the flour; because, if finely dressed, the mush would be more like dough. the coarser it is, the better it is, so that the large bran is just taken out. What a great thing is here, then, for all classes of persons, and particularly for the labourer! There may be bread made every day; you may have it hot or have it cold. There is more nutrition than you can get out of the same quantity of wheat meal; and, does not every one know, how many of those ingenious and laborious creatures, who make all the fine things that decorate our shops, live all the year round upon the meal of oats put into water! This oat-meal is, at this moment that I am writing (13th November), sold in London at *twelve shillings* for the fifty-six pounds; and corn-meal, even with the high duty of thirteen shillings and nine pence a quarter on the corn, might now be sold very well at *seven shillings* for the fifty-six pounds. There is no occasion to wait till our own crops come: there is plenty to be had from America; and plenty from *Canada*, too, without any duty that can ever be higher than half a crown a quarter. So that we may begin, in a few months' time, to live pretty nearly as well, as to the matter of bread, as the Americans themselves; except that, the *just and equitable* owners of the land and *fillers* of the seats, will make us pay a tax upon our *mush*, as they do now a tax upon our bread. It is in this state in particular, that the corn is so great a blessing as food for man; it is thus used in every house in the country: some have not the convenience of making it into bread; some may not use it in porridge; *homany* and *samp* are in use in some part of the country and not in others; but mush is used in *every house,* whether the owner be the richest or the poorest man in the country. It is eaten at the best tables, and that, almost every day; some like it hot, some cold, some with milk, some to slice it down, and eat it with meat; some like it best made with water, others with milk; but all like it in one way or another; and my belief is that the corn, even used in this one single manner, does more, as food for man, than all the wheat that is grown in the country, though the flour from that wheat is well known to be the best in the world. Will our labouring people, then, still insist upon lapping up tea-water, expensive villanous tea-water, sweetened with the not less expensive result of the sweating bodies, the aching limbs, and the bleeding backs of the Africans? will they still insist upon blowing themselves out with this costly stuff, which all the world knows has no nutrition in it? will they insist upon squandering their earnings away on the filthy and heel-swelling potatoes; when they can eat here, at a tenth part of the expense, or thereabouts, without even an oven, and without any plaguing utensil; will they continue to do this, in spite of reason; in spite of their own interest and their health; having nothing to plead in their defence, except the well-known and not very rare fact, that they *never have* eaten any of these things? People, in observing upon conduct like this, generally call it *prejudice*; a very pretty word to supply the place of *wilfulness, obstinacy, perverseness,* and brutal disregard of the dictates of reason. They have a *prejudice* against it. It unfortunately happens that they never have a *prejudice* against any food or drink, to indulge in which is ruinous. I speak of the middle class as well as of the labourers; for as to the rich, they can squander away with impunity; but in the other classes it is not only foolish, but insolent, and not only insolent, but criminal, especially in the *head* of a family, to permit this prejudice, or rather this stupid and insolent perverseness, to deprive him of the means of keeping his family well fed and in health, instead of being exposed to all the ailments arising from poverty of blood, which is occasioned by poverty of living; and to prevent him also, in many cases, from making that due future provision for his family which, if he be able, it is his duty to make. Only think of the conduct, only think of the criminal perverseness of the mother of a family, when the father is compelled to toil in one way or another from morning till night; only think of such a mother, insisting upon expending a penny upon tea slops, which have no nutrition in them; or a penny on a pound of potatoes, 'I do love a mealy potatoe, ma'am,' which has only a tenth part of nutritious matter in it; that is to say, which has only one pound of nutritious matter out of ten. Only think of such a mother; and think still more of the cowardly and the criminal conduct of the father, that thus suffers her to waste his substance, and injure his children; and all because, she does love a 'mealy potatoe' in her heart, and because she has such a *prejudice* against any sort of meal or flour, that does not proceed from wheat! But *thinking* about such a mother is not all that a husband has to do. If reasoning, if persuasion, enough to convince perverseness itself; if means of this sort be tried in vain (and they ought to be tried till all hope has expired), more effectual means ought to be resorted to; and the only means in so desperate a case is that of keeping a *tight* hand upon the *purse-strings,* which is, indeed, the only remedy; complete in all its effects, without noise and without any one disagreeable consequence. With the absence of the means of squandering, the *prejudice* is very apt to grow feebler and feebler every day; and, in this case, the porridge and the mush have all their sweetness, all their utility, and all the saving of labour and expense, at the end of about a month, generally speaking, though, in certain cases, it might last longer. Upon the disappearance of the 'prejudice', not only might the purse strings be slackened but that would be accompanied by the agreeable discovery that porridge and mush had made a considerable addition to the contents. I must, in justice to the party, observe, that I do not here speak *con amore*, as the Italians call it; for my wife has always been ready not only to imitate every good piece of management of this sort that she saw in America, but has actually made great improvements in two or three instances upon that management. She does not, indeed, abstain from the use of tea and sugar, which I am very sorry for on her own account; for the *expense*, the reader must needs think, cannot be an object with me. But she has never had any *prejudice* about any of the things proceeding from corn. She adopted the several modes of using it immediately upon seeing them; and never cried up *old England* on account of its wheat flour, as is the case with scores of perverse women whose husbands take them to America, and who, owing to the perverseness and everlasting worryings of their wives, come back and starve upon the spot from whence they started. But is it possible that we are to be told, or that we are to think for a moment, that labouring people in England, who do not see a pound of meat from month's end to month's end, will turn up their noses at food which is seen upon the tables of the most opulent people in the best fed country in the world? O no! this will not be believed; there will be for a while a little contest between the *comfortable cup of tea*, and the mess of potatoes on one side, and the porridge and mush on the other; but the contest will not last long; the love of a bellyful and the love of ease, for here the two are combined, will soon set aside the cups and saucers and the potatoe-pot, the boil of which is eternal; and I shall, I dare say, tramp about the country and see scores of families of round-faced children stuffing away upon one or the other of these articles of food.

SAMP

This, though not in such common use as porridge and **mush,**
is very much used. The corn is thumped (I do not know by what
process), as we do oats, to get the skin off it. This is put into
a pot with pork, or any other meat, and boiled, just in the
same manner as is followed by the people in the country in the
making of pea-porridge. They soak the pease over night, and
boil them with the meat the next day, and eat the porridge,
pea-shells and all. This samp is a food vastly superior to the pea;
all the pulse kinds are flatulent in their consequence; and it is
very well known that pease and beans, kidney-beans, lentiles,
tares, and, in short, all the pulse kind, if eaten, by man or
brute, to anything approaching to excess, are always dangerous, and frequently kill. I knew a farmer who was killed by
eating Windsor beans along with his men, in the harvest field;
and I had a man who died almost instantly from the eating of
kidney-beans, cooked in their ripe and dry state. There is
nothing of this sort belonging to the produce of the corn-plant;
and the samp is a great deal more nutritious as well as more
wholesome, than pea-porridge. When samp is to be made,
meat must be boiled with the cracked corn; and it must be
well boiled. American Pork is the meat in general; but the dish
might be made delightful to the most delicate palate, by boiling the skinned corn with a scrag of mutton, or a piece of lean
beef; far preferable to any thing that we have put upon the
best tables under the name of pea-soup; though the pease be
split, and though the soup be strained; for straining may take
place equally well in the case with samp, which is corn-soup
instead of pea-soup. There must be *no skins* in the samp, which
is not to be merely cracked, but the skins beaten off in the
same manner as the Dutch and the Americans beat the shells
off from buck-wheat, and as the people in the North of England
and in Scotland beat off the shells of oats before they are
ground into oat-meal. I saw a windmill in Lancashire, in which
I was surprised to see great quantities of the piths of oats

cleared from the shell. What they call pearl barley is, I suppose, prepared by the same sort of process. I was first *enlightened*; my mind made its first *march*, in this affair, on board of a little ship, called the Mary, going from Havre de Grace to New York, in the year 1792 (Good God! have I been writing ever since that time!); during which voyage, which lasted five or six and forty days, part of September, the whole of October, and part of November, the 'Good Sloop Mary', burthen ninety tons, or thereabouts, was tossed about upon the ocean, like a cork. All the fowls were dead, in somewhat the way of Tolgol's sheep. It could not be called *natural* death, indeed, for they were washed to death by the spray and the waves. But, at any rate, they almost all fell to the lot of the sailors. The turkeys did not last a week; the geese got very poor, in spite of corn; we had no pigs; and some Rouen ducks were the only things that gave us any fresh meat at all during this long and most stormy voyage. There was a Frenchman on board, named Lachaine, who, together with me and my wife, formed the passengers. The Captain, who was a *Yankee*, and whose name was Grinnell, and who was a most clever, cheerful, and obliging fellow, resorted to all the resources within his reach to furnish us with something that we could eat; and especially my wife, who, from her peculiar situation, required something other than mere pork and biscuit. One day, he said, 'I wish we could get some *samp*;' but to have samp, the skins of the grains of the corn must be beaten off. Full of contrivances, as all Yankees are, he put some corn into the bottom of a barrel, I think it was, and thumped it with a hand-spike; all done in a very clean manner; and made us some samp; the eating of which samp, gave rise to a dispute upon a question relating to the *fair sex*, or as the Irish call them, to the *heavenly part of the creation*; the nature and result of which dispute I am induced to give an account of here, for the benefit of both sexes; and particularly for the benefit of young men. Samp has ascribed to it a quality which has a tendency to produce effects precisely the opposite of those which are aimed at in the doctrines

and precepts of Parson Malthus, Peter Thimble and Carlile.*
This may be nonsense, as far as I know, but this is what people
say. The Frenchman, notwithstanding his national conceit and
pride on this subject of cookery, was delighted with the samp,
which he liked not for his own sake, but because his *intended* at
New York was so very fond of it. When a Frenchman is in love,
or when he conceits that any girl is in love with him, he takes
care to tell all the world of it; and this Frenchman had been
bragging to us, from the first moment we saw him, about his
intended at New York, who and whose sister were milliners,
living and boarding at the house of a Madame L'Epine, a
Yankee woman, who had married a Frenchman. He had told
us forty times over, that he was to be married upon his arrival;
that every thing was prepared for the nuptials; he showed my
wife trinkets and dresses that he was taking out for the occasion; he had, besides, in most elegant and curious wrought
cages, *every species of singing bird* known in Europe, not excepting
the *fauvette*, which we have not in England, and the nightingale,
so very difficult to keep in a tame state. He had all, of full age
and plumage, except the bullfinch; and not to be deficient
even here, he had a nest of young bullfinches, half fledged. He
had not excepted even the sparrow, on account, I suppose, not
so much for its *delightful song*, as for its being poetically deemed
emblematical of ardour in the affairs of love. He showed us, or
at least showed my wife, letters from his *intended*, expressive of
sentiments at which scarcely any reader will be at a loss pretty
accurately to guess. When we came to feast upon the samp, we
had Miss Hicks served up to us again, piping hot, and I, really
out of compassion for Lachaine, looking him very seriously in
the face, said, 'But, you don't think that she will wait for you,
do you?' He was a very good-humoured man; he was a furrier,
who got skins down from Canada, and carried on a traffic
with France. He was easy in his circumstances, but he was on
the wrong side of thirty; and we understood that the girl was

*A roundabout way of describing an aphrodisiac [Ed]

only about twenty. When I asked him the question, the spoon almost dropped from his hand, and his colour grew fairly red with anger. Whereupon I perceived that the business was serious; that he was really far gone; and, still preserving a serious and firm tone, I insisted that it was unreasonable to expect that a Yankee girl should wait, not only seven months, but seven days after she was in the mind. He had learned that I had been married only six months before, that my wife was then only eighteen and a half; and that she had been absent from me, and that we had had the sea between us from the age of fourteen to eighteen. Upon this ground he construed my words into an imputation against Miss Hicks; whom, he insisted that I inferred was less endowed with patience than my wife had been. I told him that no such inference was to be drawn; and that all depended upon *country*, entirely upon country; and that I knew that *Yankee* girls would not wait under such circumstances. The dispute grew very warm; and at last, in order to put an end to it, we appealed to the Captain. The Captain, who was a married man, and had a family at Boston, had uniformly taken the side of the Frenchman in all the various disputes about country and government, and other matters, with disputes about which we wiled away the time that the 'good ship Mary' kept us tossing about; but, in this case I was not afraid to appeal even to him, so confident was I of a decision in my favour. 'Now then,' said I, 'Captain Grinnell, you have heard the whole story, do you think that Miss Hicks will wait for Monsieur Lachaine, or do you not?' Both of us looked hard at the Captain, and Lachaine with manifest anxiety and fear, though he put on a smile. The Captain, clapping his two elbows upon the table, folding his hands together, and looking in a very pleasant manner, Lachaine in the face, said, 'I am sorry, Monsieur, to decide in the favour of this d-d saucy Englishman; but, I know my countrywomen; and, at that age, I know that they *will not keep.*' My wife who had taken a warm part with the Frenchman, it being a case in which the sisterhood were concerned, exclaimed 'for shame, Captain Grinnell, and you a married man too.' As to Lachaine, while he applauded this indignation of his advocate, and affected to laugh at the decision, he was manifestly stricken to the very heart. He, from that moment, dropped down into a silent and sad individual, and there was not a smile upon his face for the remainder of the voyage, which lasted another month. By this time, his singing-birds began to die; the black-bird was found dead one morning, the thrush another, another morning the lark, and the linnet, and two or three others; and so on, till all were dead except the young bullfinches, which my wife had fed, and which were fledged, and had got their fine plumage before the end of the voyage. I told the Captain (for the name of Miss Hicks was never more mentioned to Lachaine) that this dropping off of the birds was a type of the waning passion of Miss Hicks. At last the voyage ended, and we were, agreeably to an invitation a month old, to go and dine, or sup, whichever it might be, at Madame L'Epine's and to be introduced to the intended. To Mrs L'Epine's we all went; and that lady took my wife aside, even before she had got into the parlour. She soon joined us; and, pulling me aside, she whispered in my ear, 'Miss Hicks is married.' I was going to burst out; but she gave my arm a pinch, and I held my tongue. Mrs L'Epine gave us some fine oysters, fried in batter which we all gobbled up as fast as we could; and I, giving the Captain a pull and a wink, said that I must take my wife immediately down to the house of a friend who was waiting for us; for which lie I beg pardon, for Grinnell and Lachaine were the only two persons in the country that I knew even the names of. Coming out at the door, the Captain sighed out, 'Poor fellow!' shook me by the hand, and off we went, leaving Lachaine and Mrs L'Epine to their agreeable *éclaircissement*. I saw the Frenchman, a few years afterwards, at Philadelphia; he was quite an altered creature, looked to be three score and ten, and, withal, had got into poverty; and I have not the smallest doubt, that it was the cruel disappointment that he experienced that was the principal cause

of this unfavourable change. Let every young man remember
this, and particularly if he intend to have for his wife a native
of New England, New York, or New Jersey. As Grinnell said,
they will not *keep,* they are good, they are beautiful, they are
kind, they make dutiful, cleanly, and good managing wives;
they are virtuous towards their husbands, they are excellent
mothers, and are deficient in none of the duties of good
neighbourhood and hospitality: but, if, after arriving at the age
of sixteen, you once put it into their heads that you intend to
marry them, *keep* they will not

SOURCES

Introductory Observations	*Treatise on Cobbett's Corn*
Gardening	
The Situation	*The English Gardener*
Box Edging	*The English Gardener*
The Greenhouse	*The English Gardener*
Some Herbs and	
Vegetables	*The English Gardener*
except: Mustard	*Cottage Economy*
Potatoe	*Journal of a Year's Residence in the United States of America*
Garden Seeds	*Cobbett's Political Register*
Trees and Tree Planting	*The Woodlands*
Pests	*The English Gardener*
Gardeners	*The Woodlands*
A Kalendar of Work	*The English Gardener*
Farming	
Cobbett's Farm	*Cobbett's Political Register*
Wanted	*Cobbett's Political Register*
Italian Clover	*Cobbett's Political Register*
Cobbett's Sow	*Cobbett's Political Register*
Ploughing	*Treatise on Cobbett's Corn*
Wheat Growing	Introduction to *Horse Hoeing Husbandry* by Jethro Tull
The Mill	*Cobbett's Political Register*
Making Bread	*Cottage Economy*
Keeping Pigs	*Cottage Economy*
except last paragraph	*Journal of a Year's Residence in the United States of America*
Fowls	*Cottage Economy*
Pigeons	*Cottage Economy*
Rabbits	*Cottage Economy*
Goats	*Cottage Economy*
Brewing Beer	*Cottage Economy*
RURAL SPORTS	
The Game	*Rural Rides*
In Defence of Blood Sports	*Journal of a Year's Residence in the United States of America*
Single Stick	*The Life and Letters of William Cobbett* by Lewis Melville
Hare Coursing	*Rural Rides*
Hare Hunting	*Rural Rides*
Two Sweetcorn Recipes	*Treatise on Cobbett's Corn*